ZEITSCHRIFT FÜR ANGEWANDTE MATHEMATIK UND MECHANIK

INGENIEURWISSENSCHAFTLICHE FORSCHUNGSARBEITEN

BAND 62 1982 HEFT 8

AUS DEM INHALT

HAUPTAUFSÄTZE

AKADEMIE-VERLAG · BERLIN

ISSN 0044-2267 Z. angew. Math. u. Mech., Berlin **62** (1982) 8, 357—428 EVP 18,— M 34115

INHALT

Hauptaufsätze

Kleine Mitteilungen

Wir bitten, **Manuskriptsendungen** zweifach (Original und eine Kopie, sprachlich einwandfrei, Formeln mit Maschine oder in Druckschrift geschrieben) an folgende Anschrift zu richten:

Zeitschrift für Angewandte Mathematik und Mechanik, z. Hd. Herrn Prof. Dr. Günter Schmidt
Institut für Mechanik der Akademie der Wissenschaften der DDR
DDR-1199 Berlin, Rudower Chaussee 5

Zu den Arbeiten, die als Hauptaufsätze bestimmt sind, ist auf gesondertem Blatt eine Zusammenfassung von 5 bis 10 Zeilen in englischer und (möglichst) deutscher und russischer Sprache beizufügen. Ausführliche Hinweise für die Autoren, um deren strikte Berücksichtigung gebeten wird, finden sich im Anschluß an das Inhaltsverzeichnis des letzten Jahrganges.

Die Autoren erhalten von den Hauptaufsätzen 75, von den Kleinen Mitteilungen 25 Sonderdrucke ohne Berechnung, darüber hinaus weitere Sonderdrucke gegen Berechnung.

Bestellungen sind zu richten

- **in der DDR** an den Postzeitungsvertrieb unter Angabe der Kundennummer des Bestellenden oder an den AKADEMIE-VERLAG, DDR-1086 Berlin, Leipziger Str. 3—4, PSF 1233
- **im sozialistischen Ausland** an eine Buchhandlung für fremdsprachige Literatur oder an den zuständigen Postzeitungsvertrieb
- **in der BRD und Berlin(West)** an eine Buchhandlung oder an die Auslieferungsstelle KUNST UND WISSEN, Erich Bieber OHG, D-7000 Stuttgart 1, Wilhelmstraße 4—6
- **in den übrigen westeuropäischen Ländern** an eine Buchhandlung oder an die Auslieferungsstelle KUNST UND WISSEN, Erich Bieber GmbH, CH-8008 Zürich, Dufourstraße 51
- **im übrigen Ausland** an den Internationalen Buch- und Zeitschriftenhandel; den Buchexport, Volkseigener Außenhandelsbetrieb der Deutschen Demokratischen Republik, DDR-7010 Leipzig, Postfach 160; oder an den AKADEMIE-VERLAG, DDR-1086 Berlin, Leipziger Straße 3—4, PSF 1233

ZEITSCHRIFT FÜR ANGEWANDTE MATHEMATIK UND MECHANIK

Herausgeber und Chefredakteur: Prof. Dr. Günter Schmidt. Redaktion: Dr. Winfried Heinrich, Dr. Horst Weinert, Dipl.-Math. Friedhild Dudel, Bernhilde Schwertfeger, Institut für Mechanik der Akademie der Wissenschaften der DDR. Verlag: Akademie-Verlag, DDR-1086 Berlin, Leipziger Straße 3—4; Fernruf: 2236221 oder 2236229; Telex-Nr.: 114420; Bank: Staatsbank der DDR, Berlin, Kto.-Nr. 6836-26-20712. Anschrift der Redaktion: Institut für Mechanik der Akademie der Wissenschaften, DDR-1199 Berlin, Rudower Chaussee 5; Fernruf: 6743639 oder 6743643.

Veröffentlicht unter der Lizenznummer 1282 des Presseamtes beim Vorsitzenden des Ministerrates der Deutschen Demokratischen Republik. Gesamtherstellung: VEB Druckerei „Thomas Müntzer“, 5820 Bad Langensalza. Erscheinungsweise: Die Zeitschrift für Angewandte Mathematik und Mechanik erscheint monatlich. Die 12 Hefte eines Jahres einschließlich Tagungsheft bilden einen Band. Bezugspreis je Band 360,— M zuzüglich Versandspesen (Preis für die DDR 216,— M). Bezugspreis je Heft 30,— M (Preis für die DDR 18,— M). Bestellnummer dieses Heftes: 1009/62/8.

AN (EDV) 35937

ZAMM 62, 357 – 369 (1982)

H. BECKERT

The Initial Value Problem for the General Dynamic Equations in Nonlinear Elasticity Theory

In dieser Arbeit wird das Anfangswertproblem der allgemeinen dynamischen Bewegungsgleichungen eines elastischen Kontinuums gelöst. Zunächst werden die in Eulerscher Form vorliegenden Gleichungen auf eine neue Gestalt gebracht und anschließend ein einfaches Approximationsverfahren für die Lösungen über ein Differenzenverfahren in der Zeitrichtung angegeben, wobei die sich mit der Bewegung ändernden Spannungen nach der nichtlinearen Elastizitätstheorie des Verfassers [1], [2] *berechnet werden. Anschließend wird die Konvergenz der Näherungen gegen die Lösungen des obigen Problems bewiesen. Im letzten Abschnitt wird diese Theorie noch auf den Fall gedämpfter Bewegungen ausgedehnt.*

In this paper the initial value problem for the general dynamic equations in an elastic continuum is solved. At first, we introduce a new form for these equations and define a simple approximation for the solutions of our problem along a difference scheme in the time direction calculating the changing stress configurations as in the authors theory for quasistatic deformations in nonlinear elasticity [1], [2]. *The convergence of this approximation to the solution of the initial value problem above is proved. In the last chapter this theory is generalized to include damping material.*

В этой работе решается задача Коши общего динамического уравнения движения упругого континуума. Прежде всего данные уравнения, имеющие вид Эйлера, представляются в новом виде и, вслед за этим, показывается простой апроксимированный способ для их решения методом сеток в направление времени, причём эти изменяющиеся движения (при движении меняется напряжение) расчитываются на основе нелинейной теории упругости автора (1), (2). После этого доказывается сходимость приближения против решения вышеупомянутой задачи. В последнем разделе эта теория также распространяется на случай затухающего движения.

In this paper I describe a method for solving the initial value problem of the dynamic equations in nonlinear elasticity. This theory is of considerable interest for the dynamic concept of stability in elastic structures. In part A a new form for the dynamic equations in nonlinear elasticity is established. In B we introduce an approximation scheme for the solutions in A. In C the convergence of this scheme is proved in the domain of static stabiltiy, further on we show that the corresponding solutions solve our initial value problem for the general dynamic equations. In D we generalize our method to include the case of damping material. To avoid reflection from the boundary a boundary layer of high absorptivity is introduced in order to set forward our solutions.

We employ the usual notations $H_{m,2}(G)$, $\overset{0}{H}_{M,2}(G)$ for the SOBOLEV vector function spaces with scalar product $(u, v)_{m,2}$ and for the BANACH spaces $C^{m+\mu}(G)$ of $m, \mu-H$ differentiable vector functions on the domain $G \subset \mathrm{R}^3$. Volume and surface elements are denoted by $\mathrm{d}x$ and $\mathrm{d}\sigma$. Partial derivatives of order k are abbreviated by D^k. We will use the usual summation convention.

Let G_0 be an elastic body under suitable boundary conditions — say fixed boundary displacements — with a stress distribution $\sigma^0_{\mu\nu}(x^0) \subset C^\infty(G_0)$. We start with an initial impulse and initial displacements:

$$\left\{u_i^0(x^0) = \chi_i^0(x^0)\,; \qquad \frac{\partial u_i^0}{\partial t}(x^0) = \psi_i^0(x^0)\right\} \subset C^\infty(G_0)$$

$$\operatorname{supp} \chi^0(x^0)\,,\ \psi^0(x^0) \subset \operatorname{Int.} G_0 \tag{1}$$

and follow the resulting movement of the Lagrangean points x^0. In a fixed position (x, t) this movement can be described by the nonlinear dynamic equations in EULER form expressing the conservation of momentum:

$$\varrho(x, t)\,\frac{\mathrm{d}\bar{v}_i}{\mathrm{d}t}(x, t) = X_i(x) + \sigma_{ji,j}(x, t) \qquad i = 1, 2, 3, \tag{2}$$

— the force $X(x)\,\mathrm{d}x$ acting on the volume element $\mathrm{d}x$ is supposed not to change with the motion of $\mathrm{d}x$ — $\bar{v}_i(x, t)$, $\varrho(x, t)$ expressing the velocity vector and the density in the position (x, t). The corresponding initial value problem (1), (2) is a very difficult one arising from the fact that the divergence $\sigma_{ji,j}(x, t)$ of the stress tensor apparently depends on the history of the path of the Lagrangean point x^0 to x during the time interval $[0, t]$. We have a much more simpler situation in case of the NAVIER-STOKES equations, in which the corresponding terms are dependend upon the stress situation in the point x, itself. Another severe difficulty arises from the fact that in general discontinuouties will be caused by reflection from the boundary. In our theory we consider initial datas (1) with compact support in G_0 and follow the deformation as long as it arrives at the boundary ∂G_0 within the time interval, say $0 \leqq t \leqq t^+$. We construct regular solutions of the initial value problem (1) (2) along this time interval. Global regular solutions for all times cannot be expected as in the case of the linear wave equations because of the complicated interaction of changing stresses in the area of reflection. If within an arbitrary admissable small boundary layer the movement of the particles is stopped by strong damping material, our solutions can be set forward for $t > t^+$ by the same procedure as for $0 \leqq t \leqq t^+$ above. The strong regularity assumptions (1) can be reduced considerably by standard approximation methods.

A.

First some well known results on nonlinear elasticity theory: We consider an elastic body in a deformed state G_x under the stress distribution $\sigma_{\mu\nu}(x)$, $x \subset G_x$. With U_x as the elastic potential with respect to G_x this state is called static stable in the sense of E. TREFFTZ, if

$$\Delta U_x > \Delta A \tag{3}$$

holds, [7], ΔU_x is the increase of U_x under arbitrary admissable displacements $w(x)$ of the points $x \subset G_x$ and ΔA the corresponding amount of work done by the given forces. It is easily seen that the linear terms in (3) must agree on both sides. Thus, in a static stable state the first variation $V_x(w)$ of the functional $U_x - A$ is zero for all $w(x)$.

$V_x(w)$ can be deduced by mechanical principles:

$$V_x(w) = 1/2 \int_{G_x} \sigma_{\mu\nu}(x)\, \gamma_{\mu\nu}(x)\, dx - \int_{G_x} \chi_\nu^* w_\nu dx - \int_{\partial G_x} R_\nu^* w_\nu\, d\sigma = 0 \tag{4}$$

$$\gamma_{\mu\nu}(x) = \frac{\partial w_\mu}{\partial x_\nu} + \frac{\partial w_\nu}{\partial x_\mu}; \qquad 1 \leqq \mu, \nu \leqq 3\,,$$

which is equivalent to the classical conditions of equilibrium:

$$\sigma_{\mu\nu,\mu}(x) = -X_\nu^*\,, \qquad \sigma_{\mu\nu}(\sigma)\, n_\mu = R_\nu^*\,, \qquad \nu = 1, 2, 3, \tag{5}$$

with the volume forces $X^*(x)$ and boundary forces $R^*(\sigma)$. Thus, the operator $-\sigma_{\mu\nu,\mu}(x)$ must agree with the Lagrangean vector operator $L_{1,\nu}(x, 0)$ of the elastic potential U_x, see (13), (14),

$$-\sigma_{\mu,\nu,\mu}(x) = L_{1,\nu}(x, 0)\,, \tag{6}$$

so that we can write equations (2) over G_x at time t in the local form:

$$\varrho(x, t)\, \frac{dvi}{dt}(x, t) = X_i(x) - L_{1,i}(x, 0)\,. \tag{2'}$$

It follows from (3) that in a static stable state ST the second variation of the elastic potential $Q_2(w, w)$ must be positiv definite. The BANACH space $\hat{B} = \Pi C^1$ of the stress components may be divided into the regions:

ST: $Q_2(u, w, w) > 0$ of static stability,

J: $Q_2(u; w, w) \gtrless 0$ of static instability,

∂ST: $Q_2(u; w, w) \geqq 0$ but $Q(u; w^*, w^*) = 0$ on a finite dimensional subspace, [3].

From ellipticity, GARDINGS inequalitiy and a theorem of M. R. HESTENES follows over ST:

$$Q_2(u; w, w) \geqq k\, ||w||^2_{1,2}\,, \tag{7}$$

with a positiv stability constant k, [3], [6].

In [7] E. TREFFTZ derived the analytic expression of the second variation of the elastic potential U_x with respect to the deformed body G_x:

$$Q_2(w, w) = 1/2 \int_{G_x} \left[\sigma_{\mu\nu}(x)\, \frac{\partial w_i}{\partial x_\mu} \frac{\partial w_i}{\partial x_\nu} + 2a(w, w)\right] dx\,, \tag{8}$$

under the convenient assumption that the resulting stress distribution $k_{\mu\nu}(y)$ over a state arising by the admissible displacement

$$G_x \to G_y : x_i + w_i(x_i) = y_i\,,$$

can be split into the sum:

$$k_{\mu\nu}(y) = \sigma_{\mu\nu}(x) + \tau_{\mu\nu}(x)\,, \tag{9}$$

in which the additional stresses $\tau_{\mu\nu}(x)$ are to be calculated within the framework of the linear theory:

$$\tau_{\mu\nu} = \frac{\partial a}{\partial \gamma_{\mu\nu}} = T\left(\frac{\partial w}{\partial x}\right), \tag{9'}$$

$a(w, w)$ being the well known expression of the elastic potential in the linear theory .For long range deformations one has to modify the elastic constants in $a(w, w)$ to depend upon the present, or more general, past stress distributions to obtain an adequate description by means of (8).

We return to the dynamic equations (2'). Let $u(x, t - t_1)$ define an elastic movement starting from G_x at $t = t_1$:

$$G_x \to G_z : x + u(x, t_2 - t_1) = z\,; \qquad u(x, 0) = 0 \tag{10}$$

We denote by

$$U_x(v) = \int_{G_x} \Phi\left(x_j, \frac{\partial v_i}{\partial x_j}\right) dx \tag{11}$$

the elastic potential with respect to G_x and by

$$U_z(w) = \int_{G_z} \overline{\Phi}\left(z, \frac{\partial w_i}{\partial z_j}\right) dz\,, \tag{12}$$

the corresponding one with respect to G_z arising from the transformation (10) and the setting:

$$v_i(x) = u_i(x, t_2 - t_1) + w_i(z) = u_i(x, t_2 - t_1) + w_i(x + u_j(x, t_2 - t_1))\,. \tag{13}$$

From this follows:

$$\overline{\Phi}\left(z, \frac{\partial w}{\partial z}\right) = \Phi\left(x(z), \frac{\partial u_i}{\partial x_\mu} + \frac{\partial w_i}{\partial z_l}\left\{\delta_{l\mu} + \frac{\partial u_l}{\partial x_\mu}\right\}\right)\left\|\frac{dx}{dz}\right\|. \tag{14}$$

x is in $u(x, t - t_1)$ a Lagrangean variable, thus:

$$\frac{\partial u_i}{\partial t}(x, 0) = \bar{v}(x, t_1)\,; \qquad \frac{\partial^2 u_i}{\partial t^2}(x, 0) = \frac{d\bar{v}_i}{dt}(x, t_1)\,. \tag{15}$$

So we receive for the two dynamic equations (2') over G_x and G_z at time t_1 and t_2 by multiplying with

$$\xi_i(x) \subset \overset{\circ}{H}_{1,2}(G_x)\,, \qquad \eta_i(z) = \eta_i(x + u(x, t_2 - t_1)) = \xi_i(x)\,, \tag{16}$$

after integration:

$$\int_{G_x} \varrho(x) \frac{\partial^2 u_i}{\partial t^2}(x, 0)\, \xi_i(x)\, dx = (X, \xi)_0 - \int_{G_x} L_{1,i}(x, 0)\, \xi_i(x)\, dx\,, \tag{17a}$$

$$\int_{G_z} \varrho(z) \frac{\partial^2 u_i}{\partial t^2}(x, t_2 - t_1)\, \eta_i(z)\, dz = (X, \eta)_0 - \int_{G_z} L_{1,i}(z, 0)\, \eta_i(z)\, dz\,. \tag{17b}$$

Setting $p'_{i,k} = \dfrac{\partial w_i}{\partial z_k}$, $p_{i,k} = \dfrac{\partial v_i}{\partial x_k}$ in (12), (11), we receive by (14), (16) after the transformation back to G_x:

$$\int_{G_z} L_{1,i}\left(z, \frac{\partial w}{\partial z}\right)\eta_i(t)\, dz = \int_{G_z} \overline{\Phi}_{p'_{i,k}}\left(z, \frac{\partial w}{\partial z}\right)\frac{\partial \eta_i}{\partial z_k}\, dz = \int_{G_x} \Phi_{p_{i,k}}\left(x, \frac{\partial u_i}{\partial x_\mu} + \frac{\partial w_i}{\partial z_l}\left\{\delta_{l\mu} + \frac{\partial u_l}{\partial x_\mu}\right\}\right)\frac{\partial \xi_i}{\partial x_k}\, dx\,,$$

$$\int_{G_x} L_{1,i}\left(x, \frac{\partial v}{\partial x}\right)\xi_i(x)\, dx = \int_{G_x} \Phi_{p_{i,k}}\left(x, \frac{\partial v}{\partial x}\right)\frac{\partial \xi_i}{\partial x_k}\, dx\,.$$

Thus, the difference of the dynamic equations (17b), (17a) $D(t_2) - D(t_1)$ shows by conservation of mass for $v = 0$, $w = 0$, according to (2'):

$$\int_{G_x} (D_i(t_2, x) - D_i(t_1, x))\, \xi_i(x)\, dx = \int_{G_x} \varrho(x)\left\{\frac{\partial^2 u_i}{\partial t^2}(x, t_2 - t_1) - \frac{\partial^2 u_i}{\partial t^2}(x, 0)\right\}\xi_i(x)\, dx\, +$$

$$+ \int_0^1 \int_{G_x} \Phi_{p_{i,k} p_{j,s}}\left(x, t\frac{\partial u_i}{\partial x_\nu}(x, t_2 - t_1)\right)\frac{\partial u_j}{\partial x_s}(x, t_2 - t_1)\frac{\partial \xi_i}{\partial x_k}\, dx\, dt\,, \quad \forall \xi_i(x) \subset \overset{\circ}{H}_{1,2}(G_x)\,. \tag{18}$$

If the limit relations

$$\lim_{t_2 \to t_1} \frac{1}{t_2 - t_1}\frac{\partial u_j(x, t_2 - t_1)}{\partial x_s} = \frac{\partial^2 u_j}{\partial x_s \partial t}(x, t)\bigg|_{t=0}\,, \tag{19a}$$

$$\lim_{t_2 \to t_1} \frac{1}{t_2 - t_1}\left\{\frac{\partial^2 u_i}{\partial t^2}(x, t_2 - t_1) - \frac{\partial^2 u_i}{\partial t^2}(x, 0)\right\} = \frac{\partial^3 u_i}{\partial t^3}(x, t)\bigg|_{t=0}\,, \tag{19b}$$

exist, we receive after dividing (18) by $t_2 - t_1$

$$\lim_{t_2 \to t_1} \frac{D(t_2, x) - D(t_1, x)}{t_2 - t_1} = D'(t_1) = \varrho(x)\frac{\partial^3 u_i}{\partial t^3}(x, t)\bigg|_{t=0} + L_2\left(\frac{\partial u}{\partial t}(x, t)\bigg|_{t=0}\right). \tag{20}$$

$L_2 u_t$ is the Jacobian operator of the second variation (8) of the elastic potential relative to G_x. The limit system (20) is the starting point of our approach to solve the initial value problem of the dynamic equations in nonlinear elasticity theory.

B.

To study the motion starting in a deformed stable state G_0, $\{\sigma^0_{\mu\nu}(x^0)\}$, under the initial conditions (1) we introduce the following approximation scheme, similar to our method, we have used for describing quasistatic elastic deformations [1], [2]. If we subtract the equilibrium conditions (4) over G_0 at time $t = 0$ from (17b), we see that the dynamic equations (2) at time $t = 0$

$$\varrho_0(x^0) \frac{\partial^2 u_i^0}{\partial t^2}(x^0, t) = -L_{1,i}(x^0, 0), \tag{21}$$

can be substituted by

$$\varrho_0(x^0) \frac{\partial^2 u_i^0}{\partial t^2}(x^0, t) = -L_2^0(\sigma^0)\, u^0(x^0, t). \tag{21'}$$

$L_2^0(\sigma^0)$ is the Lagrangean of the second variation $Q_2^0(u, u)$ over G_0 corresponding to the stress distribution $\sigma_{\mu\nu}^0(x^0)$. Thus, dividing the time interval I: $[0T]$:

$$0 < t_1 < t_2 \dots < t_n = T, \qquad \Delta t_i = t_{i+1} - t_i,$$

we determine in the first step the vector $u_i^0(x^0, t')$ along the time interval: $0 \leqq t' \leqq t_1$ as the solution of the initial value problem of the linear wave equation (21') under the initial conditions (1) and boundary conditions $u^0(\sigma, t) = 0$, $\sigma \subset \partial G_0$. The transition:

$$G_0 \to G_1: x^0 + u^0(x^0, t_1) = x_1, \tag{22}$$

defines at time t_1 the new stress distribution

$$\sigma_{\mu\nu}^1(x^1) = \sigma_{\mu\nu}^0(x^0) + \tau_{\mu\nu}^0(x^0), \tag{23}$$

according to (8), (9') over G_1.

The starting point for the further approximation must be equation (20) for the velocity of the particles: Along the time interval $t_1 \leqq t' \leqq t_2$ the velocity of the points $x^1 \in G_1$: $u_t^1(x^1, t' - t_1) = \psi^1(x^1, t' - t_1)$ is calculated by solving the initial value problem for the linear wave equation:

$$\varrho_1(x^1) \frac{\partial^2 \psi^1}{\partial t'^2} = -L_2^1(\sigma^1)\, \psi^1(x^1, t' - t_1), \qquad \psi^1(\sigma, t' - t_1) = 0, \qquad \sigma \subset \partial G_1, \tag{24}$$

under the initial conditions:

$$u_i^1(x^1, 0) = \psi^1(x^1, 0) = u_i^0(x^0, t_1), \qquad u_{it}^1(x^1, 0) = \psi_t^1(x^1, 0) = u_{it}^0(x^0, t_1) \qquad (24'\text{a}), (24'\text{b})$$

$L_2^1(\sigma^1)$ being the Lagrangean of $Q_2^1(\sigma_{\mu\nu}^1)$, the second variation over G_1 with the stress distribution $\sigma_{\mu\nu}^1(x^1)$. The new density $\varrho_1(x^1)$ is calculated by

$$\varrho_1(x^1) = \varrho_0(x^0) \left\| \frac{dx^1}{dx^0} \right\| \sim \varrho_0(x^0) \left(1 + \sum_{l=1}^{3} \frac{\partial u_i^0}{\partial x_i^0}(x^0, t_1)\right). \tag{25}$$

The transition $G_1 \to G_2$ is now defined by

$$x^1 + u^1(x^1, t_2 - t_1) = x^1 + \int_{t_1}^{t_2} \psi^1(x^1, t' - t_1)\, dt' = x^2, \tag{26a}$$

with the new stresses $\sigma_{\mu\nu}^2(x^2)$ and density $\varrho_2(x^2)$, according to (23), (25). In the same way we proceed further. The transitions

$$G_0 \to G_1 \to \dots G_{i-1} \to G_i: x^{i-1} + \int_{t_{i-1}}^{t_i} \psi^{i-1}(x^{i-1}, t' - t_{i-1})\, dt' = x^i \tag{26$_{i-1}$}$$

with the second variations $Q_2^i(\sigma_{\mu\nu}^i)$ and corresponding stress and density distributions: $\sigma_{\mu\nu}^i(x^i)$, $\varrho_i(x^i)$ and velocities and accelerations:

$$\psi^{i-1}(x^{i-1}, t' - t_{i-1}), \qquad \dot\psi_t^{i-1}(x^{i-1}, t' - t_{i-1}),$$

being calculated as along (24), (24)', we determine inductively

$$\psi^i(x^i, t' - t_i), \qquad \psi_t^i(x^i, t' - t_i), \qquad t_i \leqq t' \leqq t_{i+1},$$

by solving the initial value problem of the wave equation:

$$\varrho_i(x^i) \frac{\partial^2 \psi^i}{\partial t'^2}(x^i, t' - t_i) = -L^i(\sigma^i)\, \psi^i(x^i, t' - t_i), \qquad \psi^i(\sigma, t' - t_i) = 0, \qquad \sigma \subset \partial G, \tag{27}$$

under initial datas:

$$\psi^i(x^i, 0) = u_i^i(x^i, 0) = u_i^{i-1}(x^{i-1}, t_i - t_{i-1}) = \psi^{i-1}(x^{i-1}, t_i - t_{i-1}). \tag{27'a}$$

$$\psi_t^i(x^i, 0) = u_{it}^i(x^i, 0) = u_{it}^{i-1}(x^{i-1}, t_i - t_{i-1}) = \psi_t^{i-1}(x^{i-1}, t_i - t_{i-1}). \tag{27'b}$$

The transition $G_i \to G_{i+1}$ at time t_i is now defined by $(26)_i$ and the new stress and density distributions over G_{i+1} by:

$$\sigma_{\mu\nu}^{i+1}(x^{i+1}) = \sigma_{\mu\nu}^i(x^i) + \tau_{\mu\nu}^i(x^i), \qquad \tau_{\mu\nu}^i(x^i) = \frac{\partial a}{\partial \gamma_{\mu\nu}^i}, \tag{28}$$

according to (9′) and

$$\varrho_{i+1}(x^{i+1}) = \varrho_i(x^i)\left(1 + \sum_k \frac{\partial u_k^i}{\partial x_k^i}(x^i, t_{i+1} - t_i)\right). \tag{29}$$

The solution of (27) can be constructed by the well known FOURIER method over the eigenvalue problems:

$$\bar{L}_2^i(\sigma^i)\, v_k^i(x^i) = 1/\varrho_i(x^i)\, L_2^i(\sigma^i)\, v_k^i(x^i) = \lambda_k^i\, v_k^i(x^i)\,, \qquad v_k^i(x^i) = 0\,, \qquad x^i \subset \partial G\,, \qquad k = 1, 2 \ldots \text{inf.}\,, \tag{30}$$

with the corresponding partial solutions:

$$v_k^i(x^i)\cos\sqrt{\lambda_k^i}(t' - t_i)\,; \qquad v_k^i(x^i)\sin\sqrt{\lambda_k^i}(t' - t_i)\,, \tag{31}$$

if, as is supposed, a stable state can be defined by the stress distribution over G_i. The system of eigenfunctions $v_k^i(x^i)$ is complete in the HILBERT space $H_{0,2}(G_i)$ with the scalar product

$$(v_k^i, v_j^i)_{0,2} = \int_{G_i} v_k^i(x^i)\, v_j^i(x^i)\, \varrho_i(x^i)\, \mathrm{d}x^i = \delta_{k,j}\,, \qquad j, k = 1, 2 \ldots \tag{32}$$

We receive the uniformly convergent expansions:

$$\psi^i(x^i) = u_t^i(x^i, 0) = \sum_k B_k^i\, v_k^i(x^i)\,, \qquad \psi_t^i(x^i) = u_{tt}^i(x^i, 0) = \sum_k A_k^i\, v_k^i(x^i)\,, \tag{33}$$

for the initial datas $\psi^i(x^i) = \psi^i(x^i, 0)$, $\psi_t^i(x^i, 0) = \psi_t^i(x^i)$, thus, the solutions of the initial value problem (27) (27′):

$$\psi^i(x^i, t' - t_i) = \sum_k \left(B_k^i \cos\sqrt{\lambda_k^i}\,(t' - t_i) + \frac{A_k^i}{\sqrt{\lambda_k^i}}\sin\sqrt{\lambda_k^i}(t' - t_i)\right) v_k^i(x^i)\,, \tag{34}$$

$$\psi_t^i(x^i, t' - t_i) = \sum_k - \sqrt{\lambda_k^i}\left(B_k^i \sin\sqrt{\lambda_k^i}(t' - t_i) + A_k^i \cos\sqrt{\lambda_k^i}(t' - t_i)\right) v_k^i(x^i)\,. \tag{35}$$

From (34), (35) follow the relations (36) for the FOURIER coefficients:

$$\begin{aligned}&(\psi_t^i(x^i, t' - t_i), v_k^i(x^i))^2 + (\psi^i(x^i, (t' - t_i), v_k^i(x^i))^2\, \lambda_k^i = \\ &(\psi_t^i(x^i, 0), v_k^i(x^i))^2 + (\psi^i(x^i, 0), v_k^i(x^i))^2\, \lambda_k^i = A_k^{i2} + B_k^{i2}\, \lambda_k^i\,, \qquad k = 1, 2 \ldots \text{inf.}\,,\end{aligned} \tag{36}$$

which will become important for our existence proof.

Because of the initial assumptions (1) the solutions (34), (35) remain in C^∞ along our approximation scheme. From the regularity point of view we are allowed to set by iteration:

$$\begin{aligned}&\bar{L}^i(\sigma^i)^3\, \psi_t^i(x^i, t' - t_i) = \bar{L}^i(\sigma^i)\, \bar{L}^i(\sigma^i)\, \bar{L}^i(\sigma^i)\, \psi_t^i(x^i, t' - t_i) = \varphi^i(x^i, t' - t_i) \\ &\bar{L}^i(\varrho^i)^3\, \bar{\psi}^i(x^i, t' - t_i) = \vartheta^i(x^i, t' - t_i)\end{aligned} \tag{37}$$

for the vector function:

$$\bar{\psi}^i(x^i, t' - t_i) = \sum_k \sqrt{\lambda_k^i}\,(\psi^i(x^i, t' - t_i), v_k^i(x^i))\, v_k^i(x^i)\,. \tag{38}$$

By summation it follows from (36)

$$\|\varphi^i(x^i, 0)\|_{0,2}^2 + \|\vartheta^i(x, 0)\|_{0,2}^2 = \|\varphi^i(x^i, t' - t_i)\|_{0,2}^2 + \|\vartheta^i(x^i, t' - t_i)\|_{0,2}^2 = P_i^2\,. \tag{39}$$

C.

To prove the convergence of our approximation in (B) we have to work in regular C^k or by embedding in suitable $H_{m,2}$ SOBOLEV spaces. In the first part we prove.

Theorem I: *There exist numbers s, P^2, A, B, C, D that the following estimates hold along a sufficient small time interval $J_s : 0 \leqq t \leqq s$:*

$$P_i^2 < P^2\,, \qquad \|\sigma^i(x^i)\|_{6,2}^2 < C\,, \tag{40a), (40b}$$

$$\|\psi_t^i(x^i, t)\|_{6,2}^2 < B\,, \qquad \|\psi^i(x^i, t)\|_{7,2}^2 < A\,, \qquad \|\varrho_i(x^i)\|_{6,2}^2 < D\,, \tag{40c), (40d), (40e}$$

for all subdivions of J_s.

For simplicity the domain bounded by the constants above shall be denoted by $\mathfrak{D}\cdot\sigma^i(x^i)$ is the vector function of the six components of the stress tensor. By the well known embedding Theorem II, we receive the estimates:

$$\|\psi_t^i(x^i, t)\|_{4+\mu} < \bar{B}\,, \qquad \|\psi^i(x^i, t)\|_{5+\mu} < \bar{A}\,, \qquad \|\sigma^i(x^i)\|_{4+\mu} < \bar{C}\,, \qquad \|\varrho_i(x^i)\|_{4+\mu} < \bar{D}\,, \qquad 0 < \mu < 1\,, \tag{40′}$$

with constants depending on A, B, C, D and the embedding constants.

Theorem II: *If the domain G fulfils a weak cone condition in the sense of* F. BROWDER [4], *then* $H_{l,p}(G) \subset \subset C^j(G + S)$ *for* $p^{-1} - (l - j)\, n^{-1} < 0$, *further:*

$$||\psi||_{C^j(G+S)} \leqq N_1 \, ||\psi||_{l,p}\,(G)\,, \qquad \psi < H_{l,p}(G)\,,$$

if G fulfils a strong cone condition in the sense of L. NIRENBERG [57], *then $D^\beta \psi$ is* HÖLDER-*continuous over $G + S$ with exponent μ, $|\beta| \leqq j$, $\mu < (l - j) - np^{-1}$.*

To prove Theorem I we introduce the constants A, B, C, D, P^2 in such a way that the initial datas belong to the domain $\mathfrak{D}$ and calculate $\varphi_0(x^0)$, $\vartheta^0(x^0)$ in $(37)_0$ to choose a suitable constant P^2 in (40). Conversely, the well known coercive equation for the iterated strongly elliptic system $(37)_0$ gives:

$$||\psi_t^0(x^0)||_{6,2}^2\,(G_0) \leqq e\,||\varphi^0(x^0)||_{0,2}^2\,(G_0)\,, \qquad ||\overline{\psi}^0(x^0)||_{6,2}^2\,(G_0) \leqq \; ||\vartheta^0(x^0)||_{0,2}^2\,(G_0)\,, \tag{41}$$

with a fixed constant e depending on the derivatives of the initial stress components and the stability constant (7). Introducing suitable positiv constants η_i we choose B to fulfil

$$||\psi_t^0(x^0)||_{6,2}^2 + ||\overline{\psi}(x^0)||_{6,2}^2 < \eta_1 + e\,[||\varphi^0(x^0)||_{0,2}^2 + ||\vartheta^0(x^0)||_{0,2}^2] < \eta_1 + eP^2 < B\,. \tag{42}$$

From the FOURIER expansion (38) follows:

$$(\overline{L}^{i4}\psi^i, \overline{L}^{i3}\psi^i)_{0,2} = \Sigma(\lambda_k^i)^7\,(\psi^i, v_k^i)^2 = (\overline{L}^{i3}\overline{\psi}^i, \overline{L}^{i3}\overline{\psi}^i)_{0,2} = ||\vartheta^i(x^i, t)||_{0,2}^2 = (\overline{L}^i\chi^i, \chi^i)_0 \geqq k\,||\chi^i||_{1,2}^2 \tag{43$_i$}$$

with:

$$\overline{L}^i(\sigma^i)^3\,\psi^i = \sum_k(\lambda_k^i)^3\,(\psi^i, v_k^i)_0\, v_k^i = \chi^i\,.$$

Thus, for $i = 0$, from well known regularity results for strongly elliptic systems:

$$||\psi^0(x^0)||_{7,2}^2\,(G_0) \leqq e_2\,||\chi^0||_{1,2}^2 \leqq e_2/k_0\,||\vartheta^0||_{0,2}^2\,(G_0) < e_2/k_0\,P_0^2 < e_2/k_0(P^2 + \eta_2) < A\,, \tag{44}$$

with a stability constant k_0 uniformly valid over $\mathfrak{D}$ and a positiv constant η_2. For simplicity, we may choose $P > 1$, $A > 1$, $B > 1$, $D > 1$.

We now follow the change of P_i^2 and the norms of $\sigma^i(x^i)$ and $\varrho_i(x^i)$ along our approximation scheme, setting:

$$\overline{L}^i(\sigma^i + \tau^i)^3\,\psi_t^i(x^i, t_{i+1} - t_i) = \overline{\varphi}^i(x^i)\,,$$

$$\big(\overline{L}^i(\sigma^i + \tau^i)^4\,\psi^i(x^i, t_{i+1} - t_i), \overline{L}^i(\sigma^i + \tau^i)^3\,\psi^i(x^i, t_{i+1} - t_i)\big)_{0,2} = ||\overline{\vartheta}^i(x^i)||_{0,2}^2 \tag{45}$$

and evaluate the transitions:

$$\varrho_i(x^i) \to \tilde{\varrho}_i(x^i) = \varrho_i(x^i)\,\big(1 + \operatorname{div} u^i(x^i, t_{i+1} - t_i)\big) \to \varrho_{i+1}(x^{i+1})\,,$$

$$||\varphi^i(x^i)||^2 \to ||\overline{\varphi}^i(x^i)||^2 \to ||\varphi^{i+1}(x^{i+1})||^2\,; \qquad ||\vartheta^i(x^i)||^2 \to ||\overline{\vartheta}^i(x^i)||^2 \to ||\vartheta^{i+1}(x^{i+1})||^2\,. \tag{46}$$

We derive majorizing differential equations for the change of these norms and the corresponding ones of the stresses and density. Because of the complexity of these equations the right sides of which are evaluated by the norms in $\mathfrak{D}$. By familiar arguments it can be proved that our approximation can never leave the domain $\mathfrak{D}$ along a sufficient small interval J_s: $Q \leqq t \leqq s$ for all its subdivisions.

Using obvious abbreviations, the following decomposition holds:

$$\overline{\varphi}^i(x^i, t^*) = \varphi^i(x^i, t^*) + 1/\varrho_i\, D(\sigma^i)\, D\, 1/\varrho_i\, D(\sigma^i)\, D\, 1/\varrho_i\, D\,(\tau^i)\, D\psi_t^i(x^i, t^*) +$$
$$+ 1/\varrho_i\, D(\tau^i)\, D\, 1/\varrho_i\, D(\sigma^i)\, D\, 1/\varrho_i\, D(\tau^i)\, D\psi_t^i(x^i, t^*) + \ldots = \varphi^i(x^i, t^*) + S_i\,, \qquad t^* = t_{i+1} - t_i\,. \tag{47}$$

After performing the differentiations in the vector differential operator S_i we see that the factors of $D^6\psi_t^i(x^i, t^*)$ contain no derivatives of $\sigma^i(x^i)$ or $\tau^i(x^i)$ in the resulting sum, the factors of $D^5\psi_t^i(x^i, t)$ at most one derivative of $\sigma^i(x^i)$ or $\tau^i(x^i)$ and the factors of $D^4\psi_t^i(x^i, t)$ contain at most a second derivative of $\sigma^i(x^i)$ or $\tau^i(x^i)$ or their first derivatives as factors only, and so on, moerover, each term in the sum S_i must have factors of derivatives of $\tau^i(x^i)$ and $\psi_t^i(x^i, t^*)$. Squaring S_i, integrating and taking into account that the additional stresses $\tau^i(x^i)$ depend linearly on the first derivatives of $u^i(x^i, t^*) = u_i$, we receive:

$$\int_{G_i} S_i^2\varrho_i\,\mathrm{d}x^i \leqq N_1 \sum_{s_0=1,\, s_1=q=r=0}^{6,6,5,4} \int_{G_i} |D^{s_0}\psi_t^i|^2\, |D^{s_1}u^i|^2 \ldots |D^{r_1}\sigma^i|^2 \ldots |D^q\varrho_i|^2 \ldots \varrho_i\,\mathrm{d}x^i \leqq$$

$$\leqq N_1\,\Sigma \int_{G_i} |D^{s_0}\psi_t^i|^2 \ldots |D^{r_1}\sigma^i|^2 \ldots \left(\int_{t_i}^{t_{i+1}} D^{s_1}\psi^i(x', t' - t_i)\,\mathrm{d}t'\right)^2 \ldots \varrho_i\,\mathrm{d}x^i \leqq$$

$$\leqq N_1\,\Sigma \int_{G_i} |D^{s_0}\psi_t^i|^2 \ldots |D^{r_1}\sigma^i|^2 \ldots \Delta t_i \int_{t_i}^{t_{i+1}} |D^{s_1}\psi^i|^2\,\mathrm{d}t' \ldots \varrho_i\,\mathrm{d}x^i\,. \tag{48}$$

N_1, N_2 ... being in the following fixed constants, According to (40), our preceding remarks, and the bounded embedding of $H_{6,2}(G_i)$ in $C^4(G_i)$, $H_{7,2}(G_i)$ in $C^5(G_i)$, at most one factor is not bounded in C^0 in each term of the integrand on the right side of (48). So we deduce the estimate:

$$\int_{G_i} S_i^2\varrho_i\,\mathrm{d}x^i \leqq (\Delta t)^2\, F_1(A, B, C, D)\,, \qquad \Delta t = \operatorname{Max} |t_{i+1} - t_i|\,. \tag{49}$$

$F_i(A, B, C, D)$, $i = 0, 1, 2, \ldots$ are polynomials in A, B, C, D or roots of such polynomials. (47) gives:

$$|||\overline{\varphi}^i(x^i, t^*)||^2 - ||\varphi^i(x^i, t^*)||^2| \leqq 2||\varphi^i||\, ||S_i|| + ||S_i||^2 \leqq (\Delta t)^2\, F_1 + \Delta t P^2 F_2\,. \tag{50}$$

As to $\vartheta^i(x^i, t)$, we deduce from (45) integrating by parts:

$$||\overline{\vartheta}^i(x^i, t^*)||^2 = \left(1/\tilde{\varrho}_i L\overline{L}^3(\sigma^i + \tau^i)\,\psi^i(x^i, t^*), \overline{L}^3(\sigma^i + \tau^i)\,\psi^i(x^i, t^*)\right) =$$
$$= Q_2^i(\sigma^i + \tau^i)\,\left(\overline{L}^3(\sigma^i + \tau^i)\,\psi^i(x^i, t^*), \overline{L}^3(\sigma^i + \tau^i)\,\psi^i(x^i, t^*)\right) + \Delta t(\,\dots), \tag{51}$$

the last term on the right can be evaluated in the same way as the second variation $Q_2^i(\sigma^i + \tau^i)\,(\overline{L}^3, \overline{L}^3)$.

With use of obvious abbreviations we receive, performing the differentiations:

$$||\overline{\vartheta}^i(x^i, t^*)||^2 - ||\vartheta^i(x^i, t^*)||^2 = \Sigma'\,(\sigma^i + \tau^i)\,\left(D\overline{L}^3(\sigma^i + \tau^i)\,\psi^i, D\overline{L}^3(\sigma^i + \tau^i)\,\psi^i\right)_0 + \Delta t(\dots)\,. \tag{52}$$

The stroke denotes that each term in the sum on the right side must contain ψ^i and τ^i or their derivatives as factors. The differential operators

$$(\sigma^i + \tau^i)\,D\overline{L}^3(\sigma^i + \tau^i)\,\psi^i(x^i, t^*)\,, \tag{53}$$

in the sum of scalar products of (52) are of type S_i in (47) with analogical properties: each factor of $D^7\psi^i(x^i, t^*)$ contains no derivative of σ^i and τ^i, the factors of $D^6\psi^i(x^i, t^*)$ at most one derivative and the factors of $D^5\psi^i(x^i, t^*)$ contain at most a second derivative of σ^i or τ^i or their first derivatives only, and so on. Analogous properties hold for the derivatives of τ^i, σ^i and ϱ_i up to order six and five. By the bounded embedding of $H_{7,2}(G_i)$ in $C^5(G_i)$, $H_{6,2}(G_i)$ in $C^4(G_i)$ at most one factor in each term of the sum defining the differential operator (53) is not bounded in $C^0(G_i)$. Thus, by the estimations of type (48) in (52):

$$\left|||\overline{\vartheta}^i(x^i, t^*)||^2 - ||\vartheta^i(x^i, t^*)||^2\right| \leqq (\Delta t)^2\,F_3 + \Delta t\,F_4\,. \tag{54}$$

Adding (54) to (50), we get finally:

$$|\overline{P}_i^2 - P_i^2| \leqq (\Delta t)^2\,F_5 + P^2F_6\,\Delta t\,. \tag{55}$$

In order to study the transitions $\overline{P}_i^2 \to P_{i+1}^2$ we have to transform $(37)_{i+1}$ and $(43)_{i+1}$ for $t' = t_{i+1}$ from G_{i+1} by $(26)_i$ back to G_i, introducing the transformation formulas for the derivatives of:

$$\psi_t^{i+1}(x^{i+1}, 0) = \psi_t^i(x^i, t_{i+1} - t_i) = \tilde{\psi}_t^i(x^i)\,, \qquad \psi^{i+1}(x^{i+1}, 0) = \psi^i(x^i, t_{i+1} - t_i) = \tilde{\psi}^i(x^i)\,,$$
$$\sigma^{i+1}(x^{i+1}, 0) = \sigma^i(x^i) + \tau^i(x^i)\,, \qquad \varrho_{i+1}(x^{i+1}) = \tilde{\varrho}_i(x^i)\,, \tag{56}$$

up to order six, seven and six, respectively.

Setting for simplicity: $x^{i+1} = y$, $x^i = x$, $u^i = u$, $x + u = y$, we receive:

$$D_x^k\tilde{\psi}^i(x) = D_y^k\psi^{i+1}(y) + \Sigma\,D_y^{s_0}\psi^{i+1}(y)\,D_x^{s_1}u\,\dots\,D_x^{s_k}u\,, \tag{57}$$

for $k = 1, 2, \dots 7$, using appropriate abbreviations. The sum runs over $1 \leqq s_0 \leqq k$, $\sum_{i=1} s_i \leqq k$, $s_0 + \text{Max}\,s_i \leqq k + 1$.

Because of our propositions (40a ... d) and Theorem II it is easily seen that at most one factor in each term of the sum (57) is not bounded in C^0, this factor must appear in the first degree. The same is true for the derivatives of $\tilde{\psi}_t^i(x)$ and the stress components according to the bounded embedding of $H_{6,2}(G_i)$ in $C^4(G_i)$. Here we must take into account that the derivatives $D^1\tau^i$, $D^2\tau^i$, ... $D^4\tau^i$ and $D^1\varrho_i$, ... $D^4\varrho_i$ are bounded in $C^0(G_i)$ so that we may assume $\sigma_{\mu\nu}^i(x^i)$ including its derivatives up to fourth order to be bounded in $C^0(G_i)$. In the resulting decomposition:

$$\varphi^{i+1}(y) = \overline{\varphi}^i(x^i) + S_i^*\,, \tag{58}$$

for:

$$1/\varrho_{i+1}\,D(\sigma^{i+1})\,D\,1/\varrho_{i+1}\,D(\sigma^{i+1})\,D\,1/\varrho_{i+1}\,D(\sigma^{i+1})\,D\psi_t^{i+1}(y) = \varphi^{i+1}(y)\,,$$

each term of the sum S_i^* contains derivatives of $u(x)$, $\sigma^{i+1}(y)$, $\varrho_{i+1}(y)$ and $\psi_t^{i+1}(y)$. As at most one factor in each term of the transformation formulas of type (57) is not bounded in C^0 according to (40), the same must be true for all terms of S_i^*. Thus, by the same conclusions as in (48), (49), (50) it follows:

$$\left|||\varrho^{i+1}(x^{i+1})||^2 - ||\overline{\varphi}^i(x^i)||^2\right| \leqq (\Delta t)^2\,F_7 + \Delta t\,F_8P^2\,. \tag{59}$$

As to the transformation of $||\vartheta^i(x^i, t_{i+1} - t_i)||^2$, we have to transform:

$$||\vartheta^{i+1}(x^{i+1}, 0)||^2 = Q_2^{i+1}\left(\overline{L}^{i+1^3}(\sigma^{i+1})\,\psi^{i+1}, \overline{L}^{i+1^3}(\sigma^{i+1})\,\psi^{i+1}\right)\,, \tag{60}$$

from G_{i+1} by $(26)_i$ back to G_i introducing the transformation formulas (57). So we get the decomposition:

$$||\vartheta^{i+1}(x^{i+1}, 0)||^2 - ||\overline{\vartheta}^i(x^i, t_{i+1} - t_i)||^2 = R\,. \tag{61}$$

To evaluate R we observe that the right side of (60) is a sum of scalar products of the type:

$$\Sigma(\sigma^{i+1}D_r\overline{L}^3(\sigma^{i+1})\,\psi^{i+1}, D_s\overline{L}^3(\sigma^{i+1})\,\psi^{i+1})_0\,. \tag{52'}$$

By splitting the differential operators

$$D_r\overline{L}^3(\sigma^{i+1})\,\psi^{i+1}\,, \tag{++}$$

it is seen that at most one factor in each term of the resulting sum is not bounded in C^0 under our propositions (40). Applying the same conclusions as in (58), (59) we see that this must remain true for the transformed expressions of type (++) also. In the resulting sum defining R each term contains derivatives of u^i. Thus, applying SCHWARZ' inequality and estimates of type (48) we finally get:

$$\left|||\vartheta^{i+1}(x^{i+1}, 0)||^2 - ||\vartheta^i(x^i, t_{i+1} - t_i)||^2\right| = R < \Delta t\,F_9 \tag{62}$$

and further with (59) and (55), (39):

$$|P^2_{i+1} - P^2_i| \leqq (\Delta t)^2 F_{10} + \Delta t\, P^2 F_{11}\,. \tag{63}$$

Similarly, we evaluate the norm $\|\sigma^i(x^i)\|_{6,2}(G_i)$ along the transitions:

$$\sigma^i(x^i) \to \sigma^i(x^i) + \tau^i(x^i) \to \sigma^{i+1}(x^{i+1})\,. \tag{64}$$

Indeed:

$$\big|\|\sigma^i + \tau^i\|_{6,2} - \|\sigma^i\|_{6,2}\big| \leqq \|\tau^i\|_{6,2} \leqq \Delta t\, F_{12} \tag{65}$$

and according to (57):

$$\big|\|\sigma^{i+1}(x^{i+1})\|^2_{6,2} - \|\sigma^i(x^i) + \tau^i(x^i)\|^2_{6,2}\big| \leqq \Delta t\, F_{15}\,. \tag{66}$$

Thus:

$$\big|\|\sigma^{i+1}(x^{i+1})\|^2_{6,2} - \|\sigma^i(x^i)\|^2_{6,2}\big| \leqq \Delta t\, F_{16}\,. \tag{67}$$

In a similar way we derive from:

$$\bar{\varrho}_i(x^i) = \varrho_i(x^i) + \varrho_i(x^i)\, \operatorname{div} u^i(x^i, t_{i+1} - t_i)\,, \tag{68}$$

$$\big|\|\varrho_{i+1}(x^{i+1})\|^2_{6,2} - \|\varrho_i(x^i)\|^2_{6,2}\big| \leqq \Delta t\, F_{17}\,. \tag{69}$$

Lemma I: *As the initial datas lie within the domain $\mathfrak{D}$, there exists by classical arguments a sufficient small interval $J_{\delta_1}: 0 \leqq t \leqq \delta_1$ that for all its subdivisions:*

$$P^2, \quad \|\sigma^i(x^i)\|^2_{6,2}, \quad \|\varrho_i(x^i)\|^2_{6,2}$$

remain in $\mathfrak{D}$.

We are not allowed to apply the coercive inequality of type (41) to use the bounds for P^2_i because the stress components are proved to belong to $C^{4+\mu}$ only. As the initial stress components $\sigma^0(x^0)$ are highly regular we transform $\psi^i_{i}(x^i, t)$ and $\psi^i(x^i, t)$ step by step along $G_i \to G_{i-1} \to G_{i-2} \to \dots G_0$ back to G_0 by $(26)_i$ setting:

$$\psi^i_i(x^i, t) = \psi^{i-1}_{i}(x^{i-1}, t) \dots = \psi^0_{i,t}(x^i, t)\,, \tag{70a}$$

$$\psi^i(x^i, t) = \psi^{i-1}_i(x^{i-1}, t) = \dots \psi^0_i(x^0, t) \tag{70b}$$

$$\overline{L}^l(\sigma^l)^3\, \psi^l_{i,t}(x^l, t) = \varphi^l_i(x^l, t)\,, \tag{71}$$

$$\|\vartheta^l_i(x^l, t)\|^2 = Q^l_2(\sigma^l)\,\big(\overline{L}^3(\sigma^l)\,\psi^l_i(x^l, t), \quad \overline{L}^3(\sigma^l)\,\psi^l_i(x^l, t)\big)\,, \qquad l = i\,, (i-1), \dots 0\,. \tag{72}$$

The norm squares:

$$\|\varphi^l_i(x^l, t)\|^2_{0,2}, \quad \|\vartheta^l_i(x^l, t)\|^2_{0,2}$$

can be evaluated along the steps above in the same way as we did in (46) ... (62). Thus, for an sufficiently small interval: $J_{\delta_2}: 0 \leqq t \leqq \delta_2 \leqq \delta_1$

$$\|\vartheta^0_i(x^0, t)\|^2_{0,2} + \|\varphi^0_i(x^0, t)\|^2_{0,2} < P^2\,, \tag{73}$$

will hold. From $(43)_0$ (44) and (41) follows:

$$\|\psi^0_i(x^0, t)\|^2_{7,2} \leqq e_2/k_0\, \|\vartheta^0_i(x^0, t)\|^2_{0,2} \leqq e_2/k_0 P^2\,, \qquad \|\psi^0_{i,t}(x^0, t)\|^2_{6,2} \leqq e\, P^2\,. \tag{74}$$

We now transform $\psi^0_i(x^0, t)$, $\psi^0_{i,t}(x^0, t)$ step by step by $(26)_i$ back to G_i. Choosing a suitable smaller interval: $J_s: 0 \leqq t \leqq s \leqq \delta_2$, if necessary, our former conclusions starting from (57) lead finally to the estimate:

$$\|\psi^i(x^i, t)\|^2_{7,2} \leqq \|\psi^0_i(x^0, t)\|^2_{7,2} + \eta_2 \leqq e_2/k_0 P^2 + \eta_2 < A\,,$$

$$\|\psi^i_i(x^i, t)\|^2_{6,2} \leqq \|\psi^0_{i,t}(x^0, t)\|^2_{6,2} + \eta_1 \leqq e P^2 + \eta_1 < B\,. \tag{75}$$

Combining this result with Lemma I we see that our approximation scheme can never leave the domain $\mathfrak{D}$ along J_s, which proves Theorem I.

In our existence proof the regularity properties (76) of the time derivatives of $\psi^i(x^i, t)$ will be used:

$$D_t \psi^i(x^i, \bar{t}) = \psi^i_i(x^i, \bar{t}) \subset C^{4+\mu}(G_i)\,, \tag{76a}$$

$$D^k_t \psi^i(x^i, \bar{t}) \subset C^{2+\mu}(G_i)\,, \qquad k = 2, 3\,; \qquad 0 < \mu < 1\,, \tag{76b}$$

$\bar{t} = t' - t_i \leqq t_{i+1} - t_i$, with global norm bounds along J_s. (76b) follows from the wave equation:

$$D^3_t \psi^i(x^i, t) = \overline{L}(\sigma^i)\, \psi^i_i(x^i, \bar{t}) = \overline{L}(\sigma^i)^{-2}\, \varphi^i(x^i, \bar{t}) \tag{77}$$

for $k = 2$ from:

$$D^2_t \psi^i(x^i, \bar{t}) = \overline{L}^i(\sigma^i)\, \psi^i(x^i, t) = \overline{L}^{i-1}(\sigma^i)\, \chi^i(x^i, \bar{t})\,. \tag{78}$$

Because of

$$\|\chi^i(x^i, t)\|^2_{1,2}\,(G_i) \leqq 1/k_0\, P^2\,, \tag{79}$$

— see (43) —

$$D^2_t \psi^i(x^i, t) \in C^{3+\mu}(G_i)\,, \tag{80}$$

Theorem I gives:

$$\|u^i(x^i, t)\|_{5+\mu}(G_i) \leqq A\, \Delta t\,. \tag{81}$$

Starting from (81) we proceed as done in our existence proof in [1] [2]. Summing up the transitions $(26)_{i-1}$ we receive for $G_0 \to G_r$: $f_k^r(x^0) = x_k^r$, $k = 1, 2, 3$
the recursion formula:

$$x_k^{r+1} = f_k^{r+1}(x^0) = f_k^r(x^0) + u_k^r(f^r(x^0), t_{r+1} - t_r)\,,$$

$$f_k^0(x^0) = x^0\,, \qquad u^r(x^r, t' - t_r) = \int_{t_r}^{t'} \psi^r(x^r, t - t_r)\, dt\,, \tag{82}$$

with:

$$|f_k^r(x^0) \leqq |x_k^0| + rA\, \Delta t\,. \tag{83}$$

To get estimates for $f_k^r(x^0)$ in $C^{5+\mu}(G_0)$ we differentiate (82) and construct exactly as in [1] systems of linear difference equations for:

$$|D^1 f_k^r(x^0)|\,, \qquad D^2 f_k^r(x^0)|\,, \ldots |D^5 f_k^r(x^0)|\,, \tag{83'}$$

which successively can be majorized by the solutions of the corresponding linear differential equations, [1], [2].
Setting:

$$f_{(n)k}(x^0, t) = f_k^r(x_0) \quad \text{for } t\colon\ t_r \leqq t \leqq t_{r+1}\,, \qquad D^\alpha f_{(n)k}(x^0, t) = D^\alpha f_k^r(x^0)\,, \qquad 1 \leqq \alpha \leqq 5\,, \tag{84}$$

— the index n relating tò the choosen subdivision of J_s — we receive sequences of uniformly bounded step functions. It is easily seen that the ,,Grenzschwankung" of the sequences (84) in the sense of C. CARATHEODORY is zero on $(G_0 + S) \times J_s$. According to a well known theorem [8], there exist suitable subsequences of (n) such that:

$$\lim f_{(n)k}(x_0, t') \rightrightarrows f_k(x^0, t')\,, \qquad \lim D^\alpha f_{(n)k}(x^0, t') \rightrightarrows D^\alpha f_k(x^0, t')\,, \qquad 1 \leqq |\alpha| \leqq 4\,,$$

$$f_k(x^0, t') \subset C^{4+\mu'}(G_0) \times C^0(J_s)\,, \qquad 0 < \mu' < 1\,. \tag{85}$$

Defining as in (84)

$$\sigma_{(n)}\mu\nu(x^0, t') = \sigma_{\mu\nu}^r(x^r) = \sigma_{\mu\nu}^r(f^r(x^0))\,; \qquad t_r \leqq t' < t_{r+1}\,. \tag{86}$$

we deduce easily from our preceding results the estimate:

$$\|\sigma_{(n)\mu\nu}(x^0, t_{r+1}) - \sigma_{(n)\mu\nu}(x^0, t_r)\|_{4+\mu'} \leqq F_{18}\, \Delta t\,. \tag{87}$$

Thus, according to the theorem above:

$$\lim_{(n')} \sigma_{(n')\mu\nu}(x^0, t') \rightrightarrows \sigma_{\mu\nu}(x^0, t')\,,$$

$$\lim_{(n')} D^\alpha \sigma_{(n')\mu\nu}(x^0, t') \rightrightarrows D^\alpha \sigma_{\mu\nu}(x^0, t')\,,$$

$$0 \leqq |\alpha| \leqq 4\,, \qquad \sigma_{\mu\nu}(x^0, t') \subset C^{4+\mu'}(G_0) \times C^0(J_s)\,, \qquad 0 < \mu' < 1\,, \tag{88}$$

along suitable subsequences (n') of (n).
By similar arguments the uniform convergence of the density function $\varrho_{(n)}(x^0, t')$, analogically defined, can be proved:

$$\lim_{(n')} \varrho_{(n')}(x^0, t') \rightrightarrows \varrho(x^0, t') \subset C^4(G^0) \times C^0(J_s)\,. \tag{89}$$

Now let us transform the $(4 + \mu')$ continuously differentiable sections of the functions $\psi^r(x^r, t)$, $\psi_t^r(x^r, t)$, $0 \leqq t = t' - t_r \leqq t_{r+1} - t_r$, $x^r \subset G + S$ by $f^r(x^0)$ back to G_0 joining them by setting:

$$\psi_{(n)}(x^0, t') = \psi^r(f_{(n)}(x^0, t_r), t' - t_r)\,, \qquad \psi_{(n)t}(x^0, t') = \psi_t^r(f_{(n)}(x^0, t_r), t' - t_r)\,. \tag{90}$$

The functions (90) are continuous over $(G_0 + S) \times J_s$ according to (91)

$$\lim_{t' \nearrow t_{r+1}} \psi_{(n)}(x^0, t') = \psi_{(n)}(x^0, t_{r+1}) = \psi^r(x^r, t_{r+1} - t_r) =$$

$$= \psi^{r+1}(x^{r+1}, 0) = \lim_{t' \searrow t_{r+1}} \psi^{r+1}(x^{r+1}, t' - t_{r+1}) = \lim_{t' \searrow t_{r+1}} \psi_{(n)}(x^0, t')\,; \tag{91}$$

the same is true for $\psi_{(n)t}(x^0, t')$ and for the derivatives:

$$D^\alpha \psi_{(n)}(x^0, t')\,, \qquad D^{\alpha'} \psi_{(n)t}(x^0, t')\,, \qquad 1 \leqq |\alpha| \leqq 4\,, \qquad 1 \leqq |\alpha'| \leqq 3\,. \tag{+}$$

According to (40'), (76), (84) $\psi_{(n)}(x^0, t')$, $\psi_{(n)t}(x^0, t')$ and its derivatives (+) are equicontinuous and bounded on $(G_0 + S) \times J_s$. Thus, for suitable subsequences (n') of the former ones

$$\lim_{(n')} \psi_{(n')}(x^0, t') \rightrightarrows \psi(x^0, t')\,, \qquad \lim_{(n')} D^\alpha \psi_{(n')}(x^0, t') \rightrightarrows D^\alpha \psi(x^0, t')\,, \qquad 1 \leqq |\alpha| \leqq 4\,,$$

$$\lim_{(n')} \psi_{(n')t}(x^0, t') \rightrightarrows \psi_t(x^0, t')\,, \qquad \lim_{(n')} D^{\alpha'} \psi_{(n')t}(x^0, t') \rightrightarrows D^{\alpha'} \psi_t(x^0, t')\,, \qquad 1 \leqq |\alpha| \leqq 3\,. \tag{92}$$

Notice that:

$$D^\alpha\psi(x^0, t') = \lim_{(n')} D^\alpha\psi_{(n')}(x^0, t') = \lim_{(n')} \int_0^{t'} D^\alpha\psi_{(n'),t}(x^0, t)\,dt + \lim D^\alpha\psi_{(n')}(x^0, 0) = D^\alpha\psi(x^0, 0) + \int_0^{t'} D^\alpha\psi_t(x^0, t)\,dt\,,$$
$$0 \leqq |\alpha| \leqq 3\,. \tag{93}$$

Let

$$-\tilde{L}^i(\sigma_{(n)\mu\nu}(x^0, t_i))\,\psi_{(n)}(x^0, t') = \varrho_{(n)}(x^0, t_i)\,\frac{\partial^2\psi_{(n)}}{\partial t'^2}(x^0, t')\,, \qquad 0 \leqq t' - t_i \leqq t_{i+1} - t_i\,, \tag{94}$$

be the differential equation (27) transformed by $x^i = f^i(x^0)$ back to G_0. We subtract the corresponding equations for $i - 1$ from (94). Regarding (87) (83) and (84), the jump of $\varrho_{i-1}\,\frac{\partial^2\psi_{(n)}}{\partial t'^2}(x^0, t')$ when crossing $t' = t_i$ can easiy be estimated by

$$\left|\varrho_i\,\frac{\partial^2\psi_{(n)}(x^0, t_i + 0)}{\partial t'^2} - \varrho_{i-1}\,\frac{\partial^2\psi_{(n)}(x^0, t_i - 0)}{\partial t'^2}\right| \leqq F_{19}\,\Delta t\,. \tag{95}$$

Thus, because of (68) and (80) there exist suitable subsequences that

$$\lim_{(n'')} \frac{\partial^2\psi_{(n)}(x^0, t')}{\partial t'^2} \rightrightarrows D_t^2\,\psi(x^0, t')\,, \tag{96}$$

hold. As $\psi_{(n'')t}(x^0, t')$ is continuous, it follows:

$$D_t^2\,\psi(x^0, t') = \psi_{tt}(x^0, t')\,, \qquad 0 \leqq t' \leqq s\,. \tag{97}$$

Obviously the function $f_k(x^0, t')$, $k = 1, 2\ 3$ defines the path of the Lagrangean point $x^0 \subset G_0$ during time interval J_s; $\psi(x^0, t)$ is the velocity and $\psi_t(x^0, t)$ its acceleration. From (82), (84), (90) follows:

$$x^s = f_{(n)}(x^0, t_s) = f_{(n)}(x^0, t_r) + \int_{t_r}^{t_s} \psi_{(n)}(x^0, t')\,dt' \tag{98}$$

and for $n \to \infty$:

$$f(x^0, t_t) = f(x^0, t_r) + \int_{t_r}^{t_s} \psi(x^0, t')\,dt'\,, \tag{99}$$

valid for all partition points t_s, by continuouity for all $t \subset J_s$. Thus;

$$\frac{\partial f}{\partial t}(x^0, t) = \psi(x^0, t) \tag{100}$$

and further, see (93):

$$\frac{\partial^2 f}{\partial t^2} = \psi_t(x^0, t)\,. \tag{101}$$

To prove (97) we note:

$$\frac{\partial^2\psi_{(n)}(x^0, t')}{\partial t'^2} = \frac{\partial^2}{\partial t^2}\psi^r(x^r, t' - t_r)\,|_{x^r = f^r(x_0)}\,, \qquad 0 \leqq t' - t_r \leqq t_{r+1} - t_r, \tag{102}$$

and by continuouity of $\psi_{(n)t}(x^0, t') = \psi_t^r(x^r, t' - t_r)$,

$$\psi_{(n_t)}(x^0, t) - \psi_{(n_t)}(x^0, \overline{\overline{t}}) = \int_{\overline{\overline{t}}}^{t} \frac{\partial^2}{\partial t^2}\psi_{(n)}(x^0, t')\,dt'\,.$$

Letting $n \to \infty$ we find:

$$\psi_t(x^0, t) - \psi_t(x^0, \overline{\overline{t}}) = \int_{\overline{\overline{t}}}^{t} D_t^2\psi(x^0, t')\,dt'\,, \tag{103}$$

which proves (97).

The deformation $G_0 \to G_t$: $f_k(x^0, t) = x_k$ is supposed to define an homeomorphism: $x_k^0 = h_k(x, t)$ along J_s. For $n \to \infty$ we get the limit equations (104) of the equations (94) at $t = t_i$ according to (96), (97):

$$-\tilde{L}(\sigma_{\mu\nu}(x^0, t_i))\,\psi(x^0, t_i) = \varrho(x^0, t_i)\,\psi_{tt}(x^0, t_i)\,. \tag{104}$$

The coefficients of (104) depend upon the partial derivatives of $h_k(x^i, t_i)$ .There exist common subsequences of all limit relations proved till now, such that (104) is valid on the set of partition points and by continuouity then for all $t \subset J_s$. Furthermore, by Theorem I (40), (96), (98), we may choose this subsequence in such a way that the systems (27) converge uniformly on the set of partition points to the system

$$-L(\sigma_{\mu\nu}(x^i)\,\hat{\psi}(x^i, 0)) = \varrho(x^i)\,\frac{\partial^2}{\partial t^2}\,\hat{\psi}(x^i, 0) \tag{105}$$

too, where along (85):

$$\psi^i(x^i) \rightrightarrows \hat{\psi}(x^i, 0) ; \qquad \frac{\partial^2 \psi^i}{\partial t^2}(x^i, 0) \rightrightarrows \frac{\partial^2}{\partial t^2}\hat{\psi}(x^i, 0) ; \qquad \varrho_i(x^i) \rightrightarrows \varrho(x^i) \tag{106}$$

hold.

The correspondence between the systems (27) and (94) shows that the system (105) can be constructed by transforming the system (104) from G_0 to G_i by $f(x^0, t_i) = x^i$. The limit $\hat{\psi}^i(x^i, 0)$ denotes the velocity of the Lagrangean point x^0 at time t_i in x^i and $\frac{\partial^2}{\partial t^2}\hat{\psi}(x^i, 0)$ is by (96), (97), (98), (99) the time derivative of the acceleration in x^i at time t_i. By continuouity the correspondence above is valid for all $t \subset J_s$. Comparing (105) with (20) we see that $D'(t) = 0$ is fulfilled for all $t \subset J_s$, thus, with (21) (21)':

Theorem III: *The functions* (85) $f(x^0, t) = x$ *describe the elastic movement of the body* G_0 *under the initial conditions* (1), *the dynamic equations* (2) *are fulfilled under the condition that the second variation of the elastic potential is calculated according to the stability theory of* E. TREFFTZ (8), (9), (9)'.

The transition rule for the stress components (9)' in the form:

$$\sigma_{(n)\mu\nu}(x^0, t_{i+1}) - \sigma_{(n)\mu\nu}(x^0, t_i) = T\left(\frac{\partial u^i}{\partial x^i_j}(x^i, t_{i+1} - t_i)\,,\right) \tag{107}$$

apparently is the approximation of the continuous one:

$$\left.\frac{d\sigma_{\mu\nu}(x^0, t_i)}{dt}\right|_{x^0 = h(x^i, t_i)} = T(\hat{\psi}^i_{x_j}(x^i_., 0))\,. \tag{108}$$

We can generalize our existence proof to cover the case in which the second variation of the elastic potential is an arbitrary positiv definite quadratic functional with corresponding transition rules:

$$\sigma^{l+1}_{\mu\nu}(x^{l+1}) = \sigma^l_{\mu\nu}(x^l) + \varphi_{\mu\nu}\left(\sigma^l_{\mu\nu}, \frac{du^l}{\partial x^i}, x^l\right), \qquad \varphi_{\mu\nu}(\sigma^l, 0, \ldots) = 0\,. \tag{109}$$

Transition rules of type (109) must be considered for long range deformations, where, clearly, the formulas (8), (9) of TREFFTZ become inadequate. In spite of this, TREFFTZ's expression (8) of the second variation of the elastic potential has general validity. One has only to modify the elastic constants in the quadratic expression of $a(u, u)$ to depend on the stress distributions to obtain a more adequate description by means of the corresponding rule (9), (9)' for longer range deformations, [2].

Our existence proof is valid as long as the movement of the particles has no contact with the boundary S of G_0 or, as is easily seen, analytically, as long as the relations:

$$\begin{aligned} &\bar{L}^i(\sigma^i)^k\, \psi^i_i(x^i, t' - t_i) = 0\,, \ \bar{L}^i(\sigma^i)\, \bar{\psi}^i(x^i, t' - t_i) = 0\,, \qquad x^i \subset \partial G_0 = S\,, \\ &\bar{L}^i(\sigma^i)^k\, \psi^i(x^i, t' - t_i) = 0\,, \qquad k = 1, 2, 3\,. \end{aligned} \tag{110}$$

are satisfied along the chains (71), (72), (76).

If the motion has reached the boundary, discontinuouities, caused by reflection, will travel inside, contrary to the linear theory where the linear wave equations under initial datas (1) have regular solutions for all times (GOURSAT's problem). To overcome these difficulties we will suppose that within a small boundary layer $\hat{G} \subset \bar{G}_0$ the movement of the particles dies down by high damping. We show that in this case our regular solution constructed can be continued regularly. This model is of common interest in cases when the discontinuouities caused by reflection die down rapidly and have practically no influence within the elastic body considered. As there is no explicit representation for the solutions of the problems (27), (27') with a variable damping coefficient as is the case with a constant one, — see D —, we introduce C^∞-functions $\gamma'(x^i, t)$ on G_i being equal to one on the inner domain $G_i - \hat{G}$, where it is supposed damping effects to be neglected. Within the boundary layer $\hat{G}$ the function $\gamma^i(x^i, t)$ and its derivatives up to arbitrary high order shall tend to zero: supp $\gamma^i \in \hat{G}$. With this damping function we define the transitions $(26)_{i-1}$; $G_{i-1} \to G_i$ by:

$$x^i = x^{i-1} + \int_{t_{i-1}}^{t_i} \gamma^i(x^i, t')\, \psi^{i-1}(x^{i-1}, t' - t_{i-1})\, dt'\,. \tag{26$^+$}$$

As a small strip around the boundary shall remain fixed, the stress and the density cannot change there, the same is true for the iterated Lagrangean operators (110). Thus, we can perform our existence proof in exactly the same way under the new conditions. Over Theorem I and the limit relations (85), (88), (89), (92), (104), (105) it follows the crucial formula:

$$D'(x^0, t) = 0\,, \qquad x^0 \subset G - \hat{G}\,, \qquad t \subset J_s\,, \tag{111}$$

as long as the path of the Lagrangean point x^0 does note nter the strip $\hat{G}$. Thus, the dynamic equations (2) are satisfied in this region. (111) is no more valid on the strip $\hat{G}$, because the limit function $\psi(x^0, t)$ defines the velocity of the point x^0 at time t in $G - \hat{G}$ only but not on $\hat{G}$. From $(26)^+$ follows easily:

$$f(x^0, t) = f(x^0, t_0) + \int_{t_0}^{t} \gamma(x^0, t')\, \psi(x^0, t')\, dt', \tag{112}$$

if we set for the damping function

$$\gamma_{(n)}(x^0, t) = \gamma^r(f^r(x^0), t'), \qquad t_{r+1} \geqq t' \geqq t_r, \qquad \gamma(x^0, t') = \lim_{n\to\infty} \gamma_{(n)}(x^0, t'), \qquad t' < J_s. \tag{113}$$

D.

We now generalize our method to include the case of damping material with a constant damping constant of the elastic body. In our approximation scheme the linear initial value problems (27), (27)' are to be replaced by the same for the equations:

$$\frac{\partial^2\psi^i}{\partial t'^2}(x^i, t'-t_i) + \frac{\alpha}{\varrho}\frac{\partial\psi^i}{\partial t}(x^i, t'-t_i) = -1/\varrho L^i(\sigma^i)\,\psi^i(x^i, t'-t_i), \tag{114}$$

with a damping constant α and *a* constant density, here supposed in D: $\varrho = 1$.

If $\{v_k^i(x^i)\}$ is the complete set of eigenfunctions of (30), we get easily the expansion (115) for the solution of the initial value problem (1), (114) in the i'th step:

$$\psi^i(x^i, t'-t_i) = \sum_k e^{-\alpha/2(t'-t_i)} B_k^i \cos\sqrt{\beta_k^i}(t'-t_i)\, v_k^i(x^i) + e^{-\alpha/2(t'-t_i)} D_k^i \sin\sqrt{\beta_k^i}(t'-t_i)\, v_k^i(x^i), \tag{115}$$

$$\psi_t^i(x^i, t'-t_i) = \sum_k -e^{-\alpha/2(t'-t_i)}\sqrt{\beta_k^i} B_k^i \sin\sqrt{\beta_k^i}\,(t'-t_i)\, v_k^i(x^i) +$$
$$+ e^{-\alpha/2(t'-t_i)}\sqrt{\beta_k^i}\, D_k^i \cos\sqrt{\beta_k^i}\,(t'-t_i)\, v_k^i(x^i) - \alpha/2\,\psi^i(x^i, t'-t_i),$$

setting, according to (33),

$$A_k\sqrt{\beta_k^i}\, D_k - \alpha/2\, B_k^i \quad \text{for} \quad \operatorname{Min}\lambda_k^i > \alpha^2/4 \quad \text{with:} \quad \sqrt{\beta_k^i} = \sqrt{\lambda_k^i - \alpha^2/4}.$$

Analogous to (36) the important relations (116) for the FOURIER-coefficients can be derived:

$$((\psi_t^i(x^i, t'-t_i), v_k^i(x^i)) + \alpha/2\,(\psi^i(x^i, t'-t_i), v_k^i(x^i)))^2 + \beta_k^i(\psi^i(x^i, t'-t_i), v_k^i(x^i))^2 =$$
$$= e^{-\alpha(t'-t_i)}((\psi_t^i(x^i, 0), v_k^i(x^i)) + \alpha/2(\psi^i(x^i, 0), v_k^i(x^i)))^2 + \beta_k^i(\psi^i(x^i, 0), v_k^i(x^i))^2, \tag{116}$$

from which follows:

$$\left(\|\varphi^i(x^i, 0) + \frac{\alpha}{2}\chi^i(x^i, 0)\|^2 + \|\vartheta^i(x^i, 0)\|^2\right) e^{-\alpha(t-t_i)} = \|\varphi^i(x^i, t'-t_i) + \frac{\alpha}{2}\chi^i(x^i, t'-t_i)\|^2 + \|\vartheta^i(x^i, t'-t_i)\|^2 =$$
$$= P_i^2(t'-t_i) \tag{117}$$

for:

$$\bar{L}^i(\sigma^i)^3\,\psi_t^i(x^i, t'-t_i) = \varphi^i(x^i, t'-t_i),$$
$$\bar{L}^i(\sigma^i)^3\,\psi^i(x^i, t'-t_i) = \chi^i(x^i, t'-t_i), \quad \text{see (43)}, \tag{118}$$
$$\bar{L}^i(\sigma^i)^3\,\tilde{\psi}^i(x^i, t'-t_i) = \vartheta(x^i, t'-t_i),$$

with:

$$\tilde{\psi}^i(x^i, t'-t_i) = \sum \sqrt{\beta_k^i}\,(\psi^i(x^i, t'-t_i), v_k^i(x^i))\, v_k^i(x^i), \tag{119}$$

according to (38). (117) gives:

$$P_i^2(0) > P_i^2(t'-t_i). \tag{120}$$

As to the evaluation of $P_i^2(0) \to P_{i+1}^2(0)$, the norm square of $\varphi^i(x^i, t'-t_i) + \alpha/2\chi^i(x^i, t'-t_i)$ can be estimated in the same way as we did before, the same is true for the estimation of $\|\vartheta^i(x^i, t'-t_i)\|^2$ along (51), (52). But here we must consider $\sqrt{\lambda_k^i}$ in (38) is to be replaced by $\sqrt{\beta_k^i}$ in (119), this leads us to the new representation:

$$\|\vartheta^i(x^i, 0)\|^2 = \hat{Q}_2^i(\bar{L}^i(\sigma^i)^3\,\psi^i(x^i, 0), \bar{L}^i(\sigma^i)^3\,\psi^i(x^i, 0)), \tag{121}$$

with the modified second variation:

$$\hat{Q}_2^i(u, u) = Q_2^i(u, u) - \alpha^2/4\;(u, u)_0, \tag{122}$$

which has the displaced eigenvalue spectrum

$$\beta_k^i = \lambda_k^i - \alpha^2/4. \tag{123}$$

With this modification we proceed estimating $\|\vartheta^i(x^i, 0)\|^2$ and P_i^2 as before. We have clearly limited the prescribed domain $\mathfrak{D}$ in (40) to belong to ST with a positiv stability constant $\hat{k}_0$ with respect to $\hat{Q}_2^i(u\;u)$ so that the estimate:

$$P^2 > \|\vartheta^i(x^i, t)\|_{0,2}^2 \geqq \hat{k}_0\,\|\chi^i(x^i, t)\|_{1,2}^2, \tag{124}$$

holds. Arrived at:

$$\|\varphi_i^0(x^0, t) + \frac{\alpha}{2}\chi_i^0(x^0, t)\|_{0,2}^2 + \|\vartheta_i^0(x^0, t)\|_{0,2}^2 < P^2, \tag{73'}$$

we determine the norm square of $\varphi_i^0(x^0, t)$ from

$$\left\| \varphi_i^0(x^0, t) + \frac{\alpha}{2} \chi_i^0(x^0, t) \right\|_{0,2} < P . \tag{125}$$

Thus,

$$\|\varphi_i^0(x^0, t)\|_{0,2} < P + \frac{\alpha}{2} \|\chi_i^0(x^0, t)\|_{1,2} , \qquad \|\vartheta_i^0(x^0, t)\|_{0,2}^2 < P^2 . \tag{126}$$

Here $\chi_i^0(x^0, t)$ has arisen from the transformation of $\psi^i(x^i, t)$ in (70) step by step back to G_0, from which follows:

$$\hat{k}_0\|\chi_i^0(x^0, t)\|_{1,2}^2 \leqq \hat{Q}_i^0(\chi_i^0, \chi_i^0) = \|\vartheta_i^0(x^0, t)\|_{0,2}^2 < \eta_2 + \|\vartheta^i(x^i, t)\|_{0,2}^2 < P^2 + \eta_2 , \quad \text{with } \eta_2 \text{ sufficiently small,} \tag{127}$$

and finally from (127):

$$\|\varphi_i^0(x^0, t)\|_{0,2}^2 < \omega P^2 , \quad \text{with} \quad \omega = \left(2 + \frac{\alpha^2}{2\hat{k}_0} + \frac{\alpha^2}{2\hat{k}_0} \eta_2\right). \tag{128}$$

The coercive inequality (41) shows:

$$\|\psi_i^0(x^0, t)\|_{6,2}^2 \leqq e\omega P^2 \tag{129}$$

and further:

$$\|\psi_i^i(x^i, t)\|_{6,2}^2 \leqq e\omega P^2 + \eta_1 . \tag{130}$$

Thus, we see that our proposition (40c) is replaced in the case of damping by:

$$\|\psi_i^i(x^i, t)\|_{6,2}^2 < e\omega P^2 + \eta_1 < B . \tag{131}$$

According to (80), (127) we get:

$$\|\psi_i^0(x^0, t)\|_{7,2}^2 \leqq e\|\chi_i^0(x^0, t)\|_{1,2}^2 \leqq e/\hat{k}_0(P^2 + \eta_2) < A , \tag{132}$$

with the same definition of A in (40d) as in C.

Now, no further modifications are necessary to accomplish our proof of Theorem III under the condition of constant damping material.

Concluding remarks:

I. In our existence proof we used the bounds in proposition (40) only, not directly higher regularity bounds. Thus, the analytic initial datas (1) in Theorem III can be replaced by datas belonging to the domain $\mathfrak{D}$, Indeed, we can approximate such datas (1)′ by analytic ones in $\mathfrak{D}$. As is easily see, the corresponding solutions along these sequences will remain in $\mathfrak{D}$ along a fixed interval J_s, by compactness, we receive then the desired solutions in Theorem III under (1)′.

II. By I, we are allowed to set forward our solution beyond J_s, as long as:

a) the norm bounds in (40) do not tend to infinity.
b) the border of stability or ellipticity is not reached.
c) the propositions on page 11 hold.

Literatur

1 BECKERT, H., Zur ersten Randwertaufgabe in der nichtlinearen Elastizitätstheorie, ZAMM **55** (1975), S. 47–58.
2 BECKERT, H., Zur zweiten und dritten Randwertaufgabe in der nichtlinearen Hyperelastizitätstheorie, ZAMM **57** (1977), S. 153 bis 164.
3 BECKERT, H., Variationsrechnung und Stabilitätstheorie, Math. Nachr. **75** (1976), S. 47–59.
4 BROWDER, F. E., On the spectral theory of elliptic differential operators I, Math. Ann **142** (1961), 23–130.
5 NIRENBERG, L., Estimates and existence of solutions of elliptic equations, Comm. on Pure and Appl. Math. **9** (1956), S. 509–530.
6 HESTENES, M. R., Applications of the theory of quadratic forms in HILBERT space to the calculus of variations, Pacific J. Math. 1 (1951), 524–571.
7 TREFFTZ., E., Zur Theorie der Stabilität des elastischen Gleichgewichts, ZAMM **13** (1933), S. 160–165.
8 CARATHEODORY, C., Reelle Funktionen Bd. I, B. G. Teubner Leipzig 1939.

Eingereicht am 1. 8. 1981

Anschrift: Prof. Dr. HERBERT BECKERT, Karl-Marx-Universität, Sektion Mathematik, DDR-7010 Leipzig, Karl-Marx-Platz, DDR

ZAMM 62, 371 –374 (1982)

F.-K. HEBEKER

Classical Solution of the Initial Value Problem for the Navier-Stokes Equations with Heat Convection

Es wird gezeigt, daß das Anfangswertproblem für kleine Zeiten eine eindeutige klassische Lösung hat. Der Beweis beruht auf dem Satz für kontraktive Abbildungen in Verbindung mit potentialtheoretischen Überlegungen. Die Regularität der Lösung wird ebenfalls diskutiert.

The initial value problem is shown to have locally in time a unique classical solution. The proof is based on the contractive mapping principle in connection with potential-theoretic resources. The solution's regularity is discussed.

Показывается, что задача Коши имеет единственное классическое решение в малом (по времени). Доказательство основано на теореме о сжимающих отображениях и на результатах из теории потенциала. Обсуждается и вопрос о регулярности решения.

1. Introduction

We study the initial value problem for the equations governing the coupled mass and heat flow of a viscous incompressible fluid in the Boussinesq approximation. In [1], CANNON and DIBENEDETTO show the existence of a local solution in L^p-spaces. To do so they develop a potential-theoretical method introduced for the Navier-Stokes equations by FABES, JONES and RIVIERE [3], which is based upon divergence-free heat potentials.

In this note we generalize the simpler method of RAUTMANN [9], which was introduced for the Navier-Stokes equations, too, and leads directly to local classical solutions. It is based upon the usual Newton and heat potentials.

2. The fixed-point equation

Let $u = u(t, x)$ be the velocity field of a flow in the n-space R^n, $n \geqq 2$, $p = p(t, x)$ the pressure function, and $\theta = \theta(t, x)$ the temperature distribution. The liquid should be incompressible and of constant viscosity $\nu > 0$ and heat conductivity $\varkappa > 0$. Corresponding to the Boussinesq approximation we assume the mass density to be constant in the differential equations with the exception of the extraneous force term (see [2]). So we consider the following *initial value problem*[1]):

$$\left.\begin{aligned} & u_t + u \cdot \nabla u - \nu \Delta u = -\nabla p + f(\theta) \\ & \nabla \cdot u = 0 \\ & u \to 0 \quad \text{when} \quad \|x\| \to \infty \\ & u = u_0 \quad \text{at} \quad t = 0; \\ & \theta_t + u \cdot \nabla \theta - \varkappa \Delta \theta = g \\ & \theta \to 0 \quad \text{when} \quad \|x\| \to \infty \\ & \theta = \theta_0 \quad \text{at} \quad t = 0 . \end{aligned}\right\} \tag{1}$$

Here f, g are known functions, u_0 (with $\nabla \cdot u_0 = 0$) and θ_0 are the given initial distributions. It is not an essential restriction to take $\nu = \varkappa = 1$.

We define the *heat potentials*

$$(\Gamma * u_0)(t, x) = \int_{R^n} \Gamma(t, x - x')\, u_0(x')\, dx' ,$$

$$(\Gamma ** u)(t, x) = \int_0^t \int_{R^n} \Gamma(t - t', x - x')\, u(t', x')\, dx'\, dt' ,$$

containing the fundamental solution

$$\Gamma(t, x) = \begin{cases} (4\pi t)^{-n/2} \cdot \exp\left(\dfrac{-x^2}{4t}\right) : & t > 0 \\ 0 \qquad : & t \leqq 0 \end{cases}$$

of the heat operator. Now (1) takes the form

$$\begin{aligned} u &= \Gamma * u_0 + \Gamma ** \left(-u \cdot \nabla u - \nabla p + f(\theta)\right), \\ \theta &= \Gamma * \theta_0 + \Gamma ** \left(-u \cdot \nabla \theta + g\right), \end{aligned} \tag{2}$$

(cp. [6, p. 265]), the boundary and incompressibility conditions are to be kept in mind.

RAUTMANN [9] introduces the singular integral operator

$$Fu = u + \nabla(\nabla \cdot \gamma * u) ,$$

---

[1]) For $x \in R^n$, $\|x\|$ denotes the euclidean and $|x|$ the maximum norm.

denoting

$$(\nabla \cdot \gamma * u)(t, x) = \int_{R^n} u(t, x') \cdot \nabla\gamma(x - x')\, dx'$$

and the fundamental solution ($c_n > 0$ constants)

$$\gamma(x) = \begin{cases} -c_2 \cdot \log \|x\| : & n = 2, \\ c_n \cdot \|x\|^{2-n} : & n = 3, 4, \dots \end{cases}$$

of the Laplacean. F projects (sufficiently regular) function spaces on their solenoidal subspaces and cancels gradients. So the pressure funktion in (2) may be eliminated, and simple transformations lead us to the following system of equations:

$$\begin{aligned} u &= \Gamma * u_0 - F\nabla \cdot \Gamma ** (u \otimes u) + F\Gamma ** f(\theta), \\ \theta &= \Gamma * \theta_0 - \nabla \cdot \Gamma ** (\theta u) + \Gamma ** g, \end{aligned} \tag{3}$$

containing the tensor $u \otimes u = u^i u^k$. The equations' right hand sides define a mapping G operating on pairs $U = (u, \theta)$. So (3) may be shortly written as a fixed point equation

$$U = GU. \tag{4}$$

By means of the contractive mapping principle the equation (4) turns out to be local in time uniquely solvable in classical function spaces which are now to be introduced.

3. Local solution of the fixed point equation

Let $J = [0, \tau]$ be a compact time interval, and $p_\lambda(x) = (1 + x^2)^\lambda$ be a weight function. Subject to (vector) functions on $J \times R^n$ we define

$$|f(t, \cdot)|_0^{(\lambda)} = \sup_{x \in R^n} |f(t, x)| \cdot p_\lambda(x);$$

$$[f(t, \cdot)]_\alpha^{(\lambda)} = \sup_{\substack{x, y \in R^n \\ x \neq y}} \{|f(t, x) - f(t, y)| \cdot \|x - y\|^{-\alpha} \cdot \min\{p_\lambda(x); p_\lambda(y)\}\},$$

$$|f(t, \cdot)|_k^{(\lambda)} = \max_{\sum j_i = j \leq k} \left| \frac{\partial^j f}{\partial x_1^{j_1} \dots \partial x_n^{j_n}}(t, \cdot) \right|_0^{(\lambda)},$$

$$|f(t, \cdot)|_{k+\alpha}^{(\lambda)} = \max\left\{ |f(t, \cdot)|_k^{(\lambda)};\ \max_{\sum j_i = j \leq k} \left[\frac{\partial^j f}{\partial x_1^{j_1} \dots \partial x_n^{j_n}}(t, \cdot) \right]_\alpha^{(\lambda)} \right\}.$$

Taking the supremum over $t \in J$ we obtain the values

$$|f|_0^{(\lambda)}, [f]_\alpha^{(\lambda)}, |f|_k^{(\lambda)}, |f|_{k+\alpha}^{(\lambda)}.$$

Let C_k denote the set of all $J \times R^n$-defined (vector) functions which are bounded and continuous with all their x-derivatives up to the k-th order. Then we need the subsets $C_k^{(\lambda)}$, $C_{k+\alpha}^{(\lambda)}$ of (vector) functions f corresponding to finite values of $|f|_k^{(\lambda)}$ and $|f|_{k+\alpha}^{(\lambda)}$, resp. Of course, time-independent (vector) functions are included in our definitions.

Now we state the following

Theorem 1: *Suppose $U_0 = (u_0, \theta_0) \in C_2^{(\lambda)}(R^n)$ with $0 < \lambda < \frac{n}{2}$ and $n \geq 3$. Let the vector function f be a Lipschitz-continuous map from a $C_0^{(\lambda)}$-neighborhood of θ_0 into one $C_0^{(\mu)}$, $\mu > \lambda$. Assume $g \in C_0^{(\lambda)}$.*

Then in a time interval $J^* = [0, \tau^*]$ there is a uniquely determined solution $U = (u, \theta) \in C_0^{(\lambda)}$ of problems (3) and (4).

The solution's regularity will be studied in Theorem 2. The proof of Theorem 1 requires a few potential-theoretic lemmas.

4. Potential-theoretic resources

Lemma 1:

(i) *For all $\lambda \geq 0$, $\alpha \in [0, 1)$, $k \in N_0$, and to each $J = [0, \tau]$ the map $\Gamma *$ is bounded and linear*

$$\Gamma *: C_{k+\alpha}^{(\lambda)}(R^n) \to C_{k+\alpha}^{(\lambda)}(J \times R^n).$$

(ii) *Assume $\lambda \geq 0$ and $U_0 \in C_2^{(\lambda)}$. Then the estimate*

$$|(\Gamma * U_0)(t, \cdot) - U_0|_0^{(\lambda)} \leq ct$$

is valid, the constant c being determined by λ and $|\Delta U_0|_0^{(\lambda)}$.

Proof: Utilize the properties of the heat potentials (e.g. [6, Chap. IV, § 1]). For (ii) use the identity $\Gamma * U_0 - U_0 = \Gamma ** (\Delta U_0)$.

Lemma 2: *For all $\lambda \geqq 0$, $\alpha \in [0, 1)$, $k \in N_0$, and to each $J = [0, \tau]$ the map Γ^{**} is bounded and linear*

$$\Gamma^{**}: C_k^{(\lambda)} \to C_{k+1+\alpha}^{(\lambda)}.$$

For any $f \in C_0^{(\lambda)}$ the estimate

$$|(\Gamma^{**} f)(t, \cdot)|_{1+\alpha}^{(\lambda)} \leqq c \int_0^t (t-t')^{-(1+\alpha)/2} |f(t', \cdot)|_0^{(\lambda)} \, dt'$$

holds, the constant c being determined by α, λ and τ. Moreover, Γ^{**} is a bounded linear map

$$\Gamma^{**}: C_{k+\alpha}^{(\lambda)} \to C_{k+2+\alpha'}^{(\lambda)}$$

for all $\alpha' \in [0, \alpha)$.

Proof: In essence the lemma is contained in [9].

Lemma 3: *Assume $n \geqq 3$ for R^n. Let $0 < \lambda < \frac{n}{2}$, $\alpha \in (0, 1)$, $k \in N_0$, and $J = [0, \tau]$. Then the mappings*

$$F\nabla: C_{k+1+\alpha}^{(\lambda)} \to C_{k+\alpha}^{(\lambda)}, \qquad F: C_{k+\alpha}^{(\lambda)} \to C_{k+\alpha}^{(\mu)}$$

are bounded and linear for all $\mu < \lambda$.

Proof: Cp. [9] for the map $F\nabla$, and [8, p. 50] for the map F.

5. Proof of Theorem 1

Our assumption guarantees the existence of numbers δ, L, $M > 0$ so that for $\theta \in C_0^{(\lambda)}$ the estimates

$$\left.\begin{aligned} |f(\theta) - f(\theta_0)|_0^{(\mu)} &\leqq L \cdot |\theta - \theta_0|_0^{(\lambda)} \\ |f(\theta)|_0^{(\mu)} &\leqq M \end{aligned}\right\} \quad \text{if} \quad |\theta - \theta_0|_0^{(\lambda)} \leqq \delta,$$

hold. Let $J = [0, \tau]$ be any time interval. To each continuous and monotone increasing function $N = N(t)$ with $N(0) = 0$ there corresponds a time interval $J_N = [0, \tau_N] \subset J$ so that

$$N(t) \leqq \frac{\delta}{2} \quad \text{when} \quad t \in J_N$$

is valid. Let A_N be the set

$$A_N = \{\varphi: J_N \times R^n \to R^{n+1} \mid |\varphi(t, \cdot) - U_0|_0^{(\lambda)} \leqq N(t), t \in J_N\}.$$

We are looking now for a function $\tilde{N}$ determining a set $A_{\tilde{N}}$ which is mapped into itself by the operator G. The lemmas imply that N has to be calculated from the integral equation

$$N(t) = ct + ct^{(1-\alpha)/2} + c \int_0^t (t-t')^{-(1+\alpha)/2} N^2(t') \, dt' \tag{5}$$

with any $\alpha \in (0, 1)$. The constant c depends on $\alpha, \lambda, \mu, \delta, M, \tau, |U_0|_2^{(\lambda)}, |g|_0^{(\lambda)}$.

The method of successive approximation shows (cp. [10]) that (5) in a time interval $\tilde{J}$ has a continuous and monotone increasing solution $\tilde{N}(t)$ which satisfies $\tilde{N}(0) = 0$. So the operator G maps the set $\tilde{A} = A_{\tilde{N}}$ into itself, if $J_{\tilde{N}} \subset \tilde{J}$ is fixed by the condition

$$\tilde{N}(t) \leqq \frac{\delta}{2} \quad \text{when} \quad t \in J_{\tilde{N}}.$$

The set $\tilde{A}$ is closed in the Banach space $C_0^{(\lambda)}$. It remains to show that G is a contraction of $\tilde{A}$ in a possibly reduced time interval. For $U = (u, \theta)$, $U' = (u', \theta') \in \tilde{A}$ the estimate

$$|(GU - GU')(t, \cdot)|_0^{(\lambda)} \leqq c \int_0^t (t-t')^{-(1+\alpha)/2} |(U - U')(t', \cdot)|_0^{(\lambda)} \, dt'$$

holds when $t \in J_{\tilde{N}}$, the constant c depending on $\alpha, \lambda, \mu, \delta, L, \tilde{A}, |U_0|_0^{(\lambda)}$.

On a sufficiently small $J^* = [0, \tau^*] \subset J_{\tilde{N}}$ the operator G is contractive. Therefore on J^* there is a uniquely determined solution U of the equation (4) so that

$$|U(t, \cdot) - U_0|_0^{(\lambda)} \leqq \tilde{N}(t) \quad \text{when} \quad t \in J^*$$

is valid. Of course, U is unique in $C_0^{(\lambda)}$.

6. The regularity of the solution

Our final aim is to study the regularity properties of the solution $U = (u, \theta)$ up to the parabolic boundary $t = 0$

Theorem 2: *The assumptions of Theorem 1 guarantee*

(i) $U \in C_{1+\alpha}^{(\lambda)}$ *for all $\alpha \in [0, 1)$.*

(ii) *To any* $\beta, \nu > 0$ *suppose* $f(\theta), g \in C_\beta^{(\nu)}$. *Then* U *in connection with*

$$p = \left(\frac{\partial}{\partial t} - \Delta\right) \{\nabla \cdot \gamma * \Gamma ** (u \cdot \nabla u - f(\theta))\}$$

is a classical solution of problem (1).

(iii) *To any* $m \in N_0$, $0 < \lambda < \frac{n}{2}$, $\alpha \in (0, 1)$, *assume* $U_0 \in C_{m+2+\alpha}^{(\lambda)}$ *and* $g \in C_{m+\alpha}^{(\lambda)}$. *Suppose that for all* $0 \leqq k \leqq m$ *and* $\beta \in (0, \alpha)$ *there exists a* $\mu > \lambda$ *so that* f *is a mapping*

$$f\colon \begin{cases} C_{k+1+\beta}^{(\lambda)} \to C_{k+\beta}^{(\mu)} \\ \theta(\cdot,\cdot) \to f(\theta(\cdot,\cdot)) \,. \end{cases}$$

Then we have

$$U \in C_{m+2+\alpha'}^{(\lambda)}, \qquad \nabla p \in C_{m+\alpha'}^{(\lambda)}, \qquad \textit{for all} \quad \alpha' \in (0, \alpha)\,.$$

(iv) *Additionally suppose that for all* $0 \leqq k \leqq m_0 \leqq \left[\frac{m}{2}\right]$ *and* $\alpha' \in (0, \alpha)$ *there exists a* $\mu > \lambda$, *so that we have*

$$\frac{\partial^k}{\partial t^k} f(\theta) \in C_{m-2k+\alpha'}^{(\mu)}, \qquad \frac{\partial^k}{\partial t^k} g \in C_{m-2k+\alpha'}^{(\lambda)}\,.$$

Then the regularity results

$$\frac{\partial^{k+1}}{\partial t^{k+1}} U \in C_{m-2k+\alpha'}^{(\lambda)}, \qquad \frac{\partial^k}{\partial t^k} \nabla p \in C_{m-2k+\alpha'}^{(\lambda)}$$

are valid for all $0 \leqq k \leqq m_0$ *and* $\alpha' \in (0, \alpha)$.

Proof: The lemmas and the well-known properties of the heat potentials imply (i) and (ii). From the lemmas we also infer

$$G\colon C_{k+\beta}^{(\lambda)} \to C_{k+1+\beta'}^{(\lambda)} \quad \text{for all} \quad 0 \leqq k \leqq m + 1\,, \quad \text{and} \quad 0 < \beta' < \beta < \alpha\,.$$

Immediately we get $U \in C_{m+2+\alpha'}^{(\lambda)}$. (3) implies

$$u = \Gamma * u_0 + v + \nabla(\nabla \cdot \gamma * v)\,,$$

with

$$v = -\Gamma ** (u \cdot \nabla u - f(\theta))\,,$$

so that u solves

$$u_t + u \cdot \nabla u - \Delta u = -\nabla p + f(\theta)$$

with the pressure function $p = \left(\frac{\partial}{\partial t} - \Delta\right)\{\nabla \cdot \gamma * \Gamma ** (u \cdot \nabla u - f(\theta))\}$. The lemmas imply $\nabla p \in C_{m+\alpha'}^{(\lambda)}$.

The final assertion (iv) is simply shown by repeated use of the relations

$$\frac{\partial}{\partial t}(\Gamma * U_0) = \Delta(\Gamma * U_0)\,, \qquad \frac{\partial}{\partial t}(\Gamma ** v) = v + \Delta(\Gamma ** v)\,.$$

References

1 CANNON, J. R.; DIBENEDETTO, E., The initial value problem for the Boussinesq equations with data in L^p, in: Approximation Methods for Navier-Stokes Problems (ed. R. RAUTMANN), Lecture Notes in Math., vol. 771, Berlin etc., Springer-Verlag, 1980, p. 129–144.

2 CHANDRASEKHAR, S., Hydrodynamic and Hydromagnetic Stability, Oxford 1961.

3 FABES, E. B.; JONES, B. F.; RIVIERE, N. M., The initial value problem for the Navier-Stokes equations with data in L^p, Arch. Rat. Mech. Analysis **45**, 222–240 (1972).

4 GAJEWSKI, H., Zur iterativen Lösung der zweidimensionalen Boussinesq-Gleichungen, ZAMM 55, 571–581 (1975).

5 KNIGHTLY, G. H., A Cauchy problem for the Navier-Stokes equations in R^n, SIAM J. Math. Analysis 3, 506–511 (1972).

6 LADYZHENSKAJA, O. A.; SOLONNIKOV, V. A.; URAL'CEVA, N. N., Linear and Quasilinear Equations of Parabolic Type, Providence (R.I.) 1968.

7 LADYZHENSKAJA, O. A., The Mathematical Theory of Viscous Incompressible Flow, New York etc. 1969.

8 MIKHLIN, S. G., Multidimensional Singular Integrals and Integral Equations, New York 1965.

9 RAUTMANN, R., On the uniqueness and regularity of the solutions of Navier-Stokes problems, Lecture Notes in Math., vol. 561, Springer-Verlag, Berlin etc., 1976, p. 378–393.

10 WALTER, W., Differential and Integral Inequalities, Springer-Verlag, Berlin etc. 1970.

Eingereicht: 10. 7. 1981

Anschrift: Dr. FRIEDRICH-KARL HEBEKER, Universität-Gesamthochschule Paderborn, Fachbereich Mathematik-Informatik, Warburger Straße 100, D-4790 Paderborn, BRD

ZAMM 62, 375–381 (1982)

G. Stoyan

On a Maximum Principle for Matrices, and on Conservation of Monotonicity. With Applications to Discretization Methods

Es werden zwei Arten von Monotonie-Eigenschaften linearer Gleichungssysteme betrachtet: Für Matrizen monotoner Art werden ein Maximumprinzip definiert und hinreichende Bedingungen für seine Gültigkeit angegeben. Für speziellere Gleichungssysteme werden Bedingungen hergeleitet, unter denen die (komponentenweise) Monotonie der rechten Seite die Monotonie der Lösung nach sich zieht. Für die untersuchten Begriffe wird eine Reihe von Beispielen betrachtet, wobei die Anwendung auf bekannte Differenzenverfahren zu einigen neuen Ergebnissen führt.

Considered are two kinds of monotonicity properties of systems of linear equations. Firstly, a maximum principle is defined for matrices of monotone kind, and sufficient conditions are given for its validity. Secondly, for more special systems, sufficient conditions are established under which (componentwise) monotonicity of the right-hand side implies monotonicity of the solution. A series of examples is considered for the investigated concepts, mainly from the context of difference schemes, leading to some new results for well-known schemes.

Рассмотрены два типа свойств монотонности для систем линейных алгебраических уравнений. Во-первых, определен принцип максимума для матриц монотонного вида и сформулированы достаточные условия для верности этого принципа. Во-вторых, для более частных систем найдены достаточные условия для того, чтобы из (покомпонентной) монотонности вектора правых частей следовала монотонность вектора решения. Для рассмотренных понятий дан ряд примеров, причем применение к хорошо известным разностным методам привело к некоторым новым результатам.

1. Introduction

For the adequate numerical solution of boundary (or initial-boundary) value problems for differential equations it is important to have discretization methods which not only approximate and are stable and effectively solvable but, moreover, reflect basic physical properties and, hence, constitute discrete physical models of the given process. Such physical properties may be expressed by additional differential inequalities satisfied by the solution (nonnegativity, monotonicity), and not any discrete approximation leads to a solution satisfying the discrete counterparts of those inequalities.

In this paper we are not interested in the various steps to discretize a given problem (see, e.g., [1] or [2]). We suppose that, finally, a system of linear equations has been obtained and consider two kinds of monotonicity properties for such systems here:

Firstly, a maximum principle is defined for matrices of monotone kind, and a sufficient condition is given for this maximum principle to hold. (We remember that difference schemes are called monotone if they satisfy the maximum principle independently on discretization parameters ([1], p. 184, p. 432).)

Secondly, for more special systems (arising from two-level discretizations of time dependent equations) sufficient conditions are given under which monotonicity is conserved, i.e. (componentwise) monotonicity of the solution vector on the old time level implies monotonicity on the new level.

Thirdly, the considered concepts and the results are illustrated and clarified by examples. Among other things it is shown that a customary difference scheme for the heat equation may satisfy the maximum principle but not conserve monotonicity, and vice versa.

2. A maximum principle for matrices of monotone kind

We shall deal with $n \times n$-matrices $A := (a_{ij})$ and with n-dimensional column vectors u, f. The identity matrix will be denoted by I, the positive (negative) part of A by $A^+(A^-)$ and the positive off-diagonal part of A by $A^{(+)}$, that is,

$$(A^{(+)})_{ij} := \begin{cases} 0, & i = j \\ \max(a_{ij}, 0), & i \neq j \end{cases}.$$

Order relations for vectors or matrices are to be understood componentwise. It follows from the context whether 0 is a number, a vector or a matrix. Further, we introduce the notations

$$N := \{1, \dots, n\}, \qquad N^0(f) := \{i \in N, f_i = 0\}, \qquad N^+(f) := \{i \in N, f_i > 0\},$$
$$e := (1, \dots, 1)^\mathsf{T}, \qquad e^j := (\delta_{ij})^\mathsf{T},$$

where δ_{ij} is the Kronecker symbol and the upper T denotes transposition.

For a matrix A of monotone kind (i.e., a regular matrix satisfying $A^{-1} \geqq 0$, see [3]–[8]) we consider the equation

$$Au = f. \tag{1}$$

Definition 1: The matrix A *satisfies the maximum principle* if (1) implies

$$u \geqq 0 \quad \text{and} \quad \max_{i \in N} u_i = \max_{j \in N+(f)} u_j > 0 \qquad \text{for all } f \geqq 0,\ f \neq 0.$$

Remarks:

1. The defined property of a matrix to satisfy the maximum principle is invariant under simultaneous permutation of columns and rows, since this is equivalent to simultaneous permutation of the components of u and f. But the property may be lost if transponing the matrix.

2. The "physical" sense of Def. 1 is obvious: A "system" fulfilling the maximum principle shows the maximal response in one of those of its components where there is a nonzero influence.

3. Let be (1) obtained by discretization (be this by difference or finite element methods) of some boundary value problem for a linear (partial) differential equation. Let the boundary conditions be included into (1) and correspond to the indices $N_1 \subseteq N$ ($N_1 \neq \emptyset$). Further, let be $f \geqq 0$, $f \neq 0$, $N^+(f) \subseteq N_1$, i.e., the right-hand side of the differential equation is zero and the right-hand sides of the boundary conditions (in case of first kind conditions: the boundary values) are nonnegative. If A is of monotone kind and satisfies the maximum principle, then the solution u is nonnegative and takes on its maximum for $i \in N^+(f) \subseteq N_1$, that is, on the boundary.

There is no difference in treating initial-boundary value problems or letting N_1 be some arbitrary subset of N.

Further results (e.g., for right-hand sides f not restricted to $f \geqq 0$) may be obtained in the way usual for problems of monotone kind, see [1], p. 247/248 or [4].

4. A necessary (but not sufficient) condition for A to satisfy the maximum principle is the dominance of the diagonal elements of A^{-1} over the elements of the corresponding columns. (To see this, take $f = e^j$.) In other words, the Green's function G_{ij} of the corresponding discretizations is maximal for $i = j$.

5. In [9], [1], p. 245, and [10] maximum principles for discretizations of linear boundary value problems have been defined and investigated. All these maximum principles (as well as the one defined here) are true if the matrix is of generalized nonnegative type in the sense of [10], p. 343. [10] allows for more general cases than [9], [1], and gives necessary and sufficient conditions for the maximum principle to hold.

The above maximum principle does not assume a structure of the matrix (connected with prescription of first kind boundary conditions) like [10] and gives more information (comp. Remark 3 to Def. 1). But for that, in turn, the sufficient conditions of Theorem 2 in [10] (i.e., the maximum principle [10] holds if A is of monotone kind and satisfies $Ae \geqq 0$) must be strengthened as is shown by Theorem 1 and Remark 3 below.

Theorem 1: *Let be $A + \varepsilon I$ of monotone kind for $\varepsilon \in [0, \varepsilon^0]$ with some positive number ε^0. Then A satisfies the maximum principle if*

$$(A - A^{(+)})\, e \geqq 0 \tag{2}$$

is true.

Proof: Along with (1), $f \geqq 0$, $f \neq 0$, we consider the "regularized" equation

$$(A + \varepsilon I)\, u^\varepsilon = f, \qquad \varepsilon \in [0, \varepsilon^0]. \tag{3}$$

Both (1) and (3) have unique, nonnegative solutions which are not identically zero due to $f \neq 0$. We may assume $N^0(f) \neq 0$. (Otherwise, the maximum principle holds trivially.) Suppose now that all matrices $A + \varepsilon I$, $\varepsilon \in (0, \varepsilon^0]$, satisfy the maximum principle, i.e.,

$$\max_{i \in N} u_i^\varepsilon = \max_{j \in N^+(f)} u_j^\varepsilon > 0. \tag{4}$$

Since $(A + \varepsilon I)^{-1}$ exists by assumption for all $\varepsilon \in [0, \varepsilon^0]$, it is continuous in ε on $[0, \varepsilon^0]$. Hence u^ε is continuous in ε and (4) holds for $\varepsilon = 0$, too.

Next, on the contrary, suppose that $A + \varepsilon I$ for some $\varepsilon \in (0, \varepsilon^0]$ does not satisfy the maximum principle, then

$$\max_{i \in N} u_i^\varepsilon = \max_{j \in N^0(f)} u_j^\varepsilon > 0.$$

Therefore, if u_j^ε is maximal, then $\varepsilon u_j^\varepsilon > 0$ and from (1)

$$(a_{jj} + \varepsilon)\, u_j^\varepsilon = -\sum_{\substack{j \neq k \\ a_{jk} < 0}}{}' a_{jk} u_k^\varepsilon - \sum_{\substack{j \neq k \\ a_{jk} \geqq 0}}{}' a_{jk} u_k^\varepsilon \leqq u_j^\varepsilon \Big(- \sum_{\substack{j \neq k \\ a_{jk} < 0}}{}' a_{jk} \Big) \leqq u_j^\varepsilon a_{jj},$$

where $u^\varepsilon \geqq 0$ and (2) have been used. Subtraction of $a_{jj} u_j^\varepsilon$ on both sides of the inequality gives $\varepsilon u_j^\varepsilon \leqq 0$, a contradiction.

Remarks:

1. If A is regular then so is $A + \varepsilon I$ for $\varepsilon < \|A^{-1}\|^{-1}$ (see, e.g., [7], p. 45). If, moreover, $A^{-1} > 0$, then $(A + \varepsilon I)^{-1} \geqq 0$ for sufficiently small ε, by continuity of $(A + \varepsilon I)^{-1}$. Hence, only (2) remains to be checked now, which is easily done.

There are well-known classes of matrices A with $A^{-1} > 0$: symmetric positive definite matrices satisfying $a_{ij} < 0$, $i \neq j$ [11], irreducible M-matrices (see [7], p. 53/54), matrices A for which there exists a matrix B of monotone kind and a (strictly) positive vector z with $B \geqq A$, $Az \geqq 0$, $\neq 0$ [5]. For the matrices A described by Theorem 3 in [12] (if in relation (4), p. 229, strict inequality holds) there exists also an $\varepsilon^0(A)$ with the properties required in Theorem 1.

2. If A is an M-matrix then $A^{(+)} = 0$, moreover, the argumentation of Remark 1 is superfluous since in that case $A + \varepsilon I$ is an M-matrix, too (see [7], p. 54/55). Now Theorem 1 reads: A *diagonally dominant M-matrix satisfies the maximum principle.*

3. There exist matrices A for which $A + \varepsilon I$ is of monotone kind for $\varepsilon \in [0, \varepsilon^0]$ for some $\varepsilon^0 > 0$, which fulfil, instead of (2), the weaker condition $Ae \geqq 0$ as in [10], but which do not satisfy the maximum principle. Also, there exist matrices A for which $A + \varepsilon I$, $\varepsilon \geqq 0$, is of monotone kind for $\varepsilon = 0$ only, and which satisfy (2) but not the maximum principle. See Examples b and c in Point 4.

3. Conservation of monotonicity

Two-level (or in other terminology: one-step) methods for initial-boundary value problems can be written in the following normal form (comp. [1], p. 348):

$$B_j \frac{y^j - y^{j-1}}{\tau_j} + A_j y^{j-1} = \varphi^j, \quad y^0 = \varphi^0, \quad j = 1, 2, \ldots \tag{5}$$

Here τ_j denotes the time step, y and φ are n-vectors, B and A are $n \times n$-matrices. y^j and φ^0 are called the *solution on the time level j and the initial value, respectively.* If the B_j are regular then all y^j can be determined from (5) step by step.

We suppose (different from [1]) boundary conditions (if posed) to be included into (5). For example, if there are given first kind conditions for some y_i^j (i.e., for y_i^j a certain value is prescribed) then we may assume the i-th rows of B_j and A_j to be equal to the i-th rows of I and 0, respectively. The "boundary" value is then determined by φ_i^j and y_i^{j-1}.

Definition 2: The two-level method (5), where $\varphi^j = 0$, $j = 1, 2, \ldots$, is said to *conserve monotonicity* if, for all j^0, the following conditions hold:

1. If the relations

$$y_1^j \leqq y_2^j \leqq \ldots \leqq y_n^j \tag{6}$$

are true for $j = j^0$, then (5) implies (6) to hold for $j = j^0 + 1$, too.

2. If

$$y_1^j \geqq y_2^j \geqq \ldots \geqq y_n^j \tag{7}$$

is true for $j = j^0$, then (5) implies (7) to be valid for $j = j^0 + 1$, too.

Remark: Already by its definition, conservation of monotonicity is a concept restricted to spatially one-dimensional problems. But it seems possible to consider, for a class of manydimensional equations, a similar property of conservation of coordinatewise monotonicity in combination with certain ADI-methods ([13], [14]).

The problem of conservation of monotonicity in the sense of Def. 2 has earlier been investigated in [15] for some methods (5) corresponding to difference schemes for constant coefficient initial value problems posed on the infinite line.

We start our investigation with a rather general condition.

Lemma 1: *Let be*

$$(B_j e)_i =: d_i^j > 0 \quad \textit{and} \quad A_j e = 0, \quad j = 1, 2, \ldots$$

Then (5) *conserves monotonicity if all matrices T_j^S are of monotone kind, where*

$$T_j^S := \begin{pmatrix} I & 0 \\ -C_j^S & B_j^S \end{pmatrix}, \quad B_j^S := S^{-1} D_j^{-1} B_j S, \quad C_j^S := S^{-1} D_j^{-1} C_j S,$$

$$D_j := \operatorname{diag}(d_i^j), \quad C_j := B_j - \tau_j A_j, \quad S := \begin{pmatrix} 1 & & & & 0 \\ \cdot & 1 & & & \\ \cdot & & \cdot & & \\ \cdot & & & \cdot & \\ 1 & \cdot & \cdot & \cdot & 1 \end{pmatrix}.$$

Proof: We observe, firstly, that the conditions imply that B_j is regular and, therefore, (5) is solvable. Obviously, it suffices to consider the case $j^0 = 0$. We make also the simple remark that every vector y^0 satisfying (6), $j = 0$, can be written as

$$y_0 = S v^0, \tag{8}$$

where $v_i^0 = (S^{-1} y^0)_i = y_i^0 - y_{i-1}^0 \geqq 0$, $i = 2, \ldots, n$, and $v_1^0 = y_1^0$. Conversely, if $v_i^0 \geqq 0$ in (8), $i = 2, \ldots, n$, then (6) holds for $j = 0$. Next, due to $A_j e = 0$ and $\varphi^j = 0$, $j \geqq 1$, we may add to φ^0, y^0 and y^1 some multiple of the vector e without changing the equations (5). That is, if (6) is true, then in the representation (8) we may assume $v_1^0 \geqq 0$ and hence $v^0 \geqq 0$. Likewise, if (7) is satisfied for $j = 0$, we may multiply φ^0, y^0 and y^1 by -1, add an appropriate multiple of e and arrive once more at (8) with $v^0 \geqq 0$.

We represent now y^1 in the same manner, $y^1 = S v^1$, and, rewriting (5) in the equivalent form

$$T_j \begin{pmatrix} y^{j-1} \\ y^j \end{pmatrix} = \begin{pmatrix} y^{j-1} \\ 0 \end{pmatrix}, \quad T_j := \begin{pmatrix} I & 0 \\ -C_j & B_j \end{pmatrix}, \tag{9}$$

we obtain for $j = 1$

$$T_1 \begin{pmatrix} S & 0 \\ 0 & S \end{pmatrix} \begin{pmatrix} v^0 \\ v^1 \end{pmatrix} = \begin{pmatrix} Sv^0 \\ 0 \end{pmatrix}.$$

Multiplying this equation from the left first by $\begin{pmatrix} I & 0 \\ 0 & D_1^{-1} \end{pmatrix}$ and then by $\begin{pmatrix} S^{-1} & 0 \\ 0 & S^{-1} \end{pmatrix}$, we get

$$T_1^S \begin{pmatrix} v^0 \\ v_1 \end{pmatrix} = \begin{pmatrix} v^0 \\ 0 \end{pmatrix}.$$

Since the right-hand side here is nonnegative and T_1^S is of monotone kind, we find $v^1 \geqq 0$ and the lemma follows.

Remarks:

1. As Def. 2 shows, any monotonicity conserving two-level method possesses a solution y^j of the form $c_j e$ in case $\varphi^0 = e$, $\varphi^j = 0$, $j \geqq 1$. If, moreover, constant first kind boundary conditions are prescribed on all time levels, then $y^j = e$. Under these circumstances, the condition $A_j e = 0$ is seen to be necessary for conservation of monotonicity by (5). Changing to other types of boundary conditions, the unchanged ("inner") equations require $A_j e = 0$ to remain true.

2. If the matrices $B_j = B$ and $A_j = A$ do not depend on j, if B is (strongly) diagonally dominant, B^S a matrix of monotone kind and $Ae = 0$, then the condition on T^S is *necessary*:

Assuming $v^1 \geqq 0$, for arbitrary $v^0 \geqq 0$ we have, necessarily, $(B^S)^{-1} C^S \geqq 0$. Hence T^S is of monotone kind.

3. The proof of Lemma 1 shows also that $y_1^{j+1} \geqq y_1^j$ (or $y_1^{j+1} \leqq y_1^j$) if (6) (or (7)) is true: We may assume $v_1^0 = 0$ in (8) and obtain $v^1 \geqq 0$ as above, i.e., $v_1^1 \geqq 0$. Similarly, by symmetry of the concept of conservation of monotonicity, $y_n^{j+1} \leqq y_n^j$ ($y_n^{j+1} \geqq y_n^j$) if (6) ((7)) holds. In other words, if y^0 is monotone, then for a monotonicity conserving method (5) where $\varphi^j = 0$, $j \geqq 1$, not only will be y^j monotone for all j, but, moreover, $|y_1^j - y_n^j|$ will not increase with j.

Using Lemma 1, for band matrices A_j and B_j rather simple sufficient conditions may be formulated which assure (5) to conserve monotonicity. We shall show this for tridiagonal matrices, i.e., matrices which have nonzero elements on the main diagonal and the two adjacent diagonals only.

First we consider properties of the matrices occurring in the block matrix T_j^S of Lemma 1.

Lemma 2: *Let be A a tridiagonal matrix with row sum 1. Then $S^{-1}AS =: (\alpha_{ij})$ is tridiagonal, too, and*

$$\alpha_{ii-1} = a_{i-1i-2}\,, \qquad \alpha_{ii} = 1 - a_{ii-1} - a_{i-1i}\,, \qquad \alpha_{ii+1} = a_{ii+1}\,,$$

$i = 1, \dots, n$ (*where elements with indices k not fitting $1 \leqq k \leqq n$ are understood to be zero*).

The proof follows by direct calculation. (The equality of the row sums is seen to be important in this calculation, and the multiplication by D_j^{-1} in Lemma 1 prepares to meet this requirement.)

We observe that the column sums of $S^{-1}AS$ are 1 except for the n-th column where the sum is $1 - a_{nn-1} = a_{nn}$.

Lemma 3: *Let be A a tridiagonal M-matrix with row sum 1. Then $S^{-1}AS$ is an M-matrix, too.*

Proof: By Lemma 2, A and $S^{-1}AS$ have, simultaneously, nonpositive off-diagonal elements. Since all column sums of the matrix $S^{-1}AS$ are positive, it is an M-matrix.

Combining these lemmas, we will prove the following result.

Theorem 2: *Let A_j and B_j be tridiagonal matrices fulfilling the following conditions for all j:*

1. *B_j is an M-matrix with positive row sums, $(B_j e)_i =: d_i^j > 0$.*
2. *$A_j e = 0$.*
3. *The matrix $C_j = (c_{ik}^j) := B_j - \tau_j A_j$ has nonnegative off-diagonal elements satisfying*

$$\frac{c_{ii-1}^j}{d_i^j} + \frac{c_{i-1i}^j}{d_{i-1}^j} \leqq 1\,, \qquad i = 2, \dots, n\,. \tag{10}$$

Then (5) *conserves monotonicity.*

Proof: Under the given conditions, the assumptions of Lemma 1 are fulfilled. By Lemma 3, B_j^S is an M-matrix. Hence, T_j^S will be of monotone kind if $(B_j^S)^{-1}C_j^S \geqq 0$. Conditions 1 and 2 show $D_j^{-1}C_j$ to have all row sums equal 1. Now condition 3 and Lemma 2 ensure $S^{-1}D_j^{-1}C_jS = C_j^S \geqq 0$, and the assertion follows.

Remarks:

1. The conditions of Theorem 2 imply that the T_j^S are M-matrices. For a given special two-level discretization (5) one may try to use Theorem 3 in [12] to obtain weaker conditions.

2. The theorem can be generalized to full matrices, for instance, if the off-diagonal elements of $D_j^{-1}C_j$ and the moduli of the off-diagonal elements of $D_j^{-1}B_j$ do not decrease in their columns in direction to the main diagonal, and if there hold inequalities similar to (10).

4. Examples

a) An $n \times n$-matrix fulfilling the maximum principle by Def. 1 is

$$A_n := \frac{1}{\delta\Delta}\,[-(a_1, \dots, a_n) + \Delta I]\,,$$

where

$$\Delta := \delta + \sum_1^n a_i\,, \qquad \delta > 0\,, \qquad a_i \geqq 0\,, \qquad i = 1, \dots, n\,,$$

and where $(a_1, \dots, a_n)$ denotes the matrix the i-th column of which has all elements equal to a_i.

In fact,

$$A_n^{-1} = (a_1, \dots, a_n) + \delta I \geqq 0\,, \tag{11}$$

and the conditions of Theorem 1 are satisfied: (2) holds, and $(A_n + \varepsilon I)^{-1}$ is nonnegative and given by a formula similar to (11). But, as (11) shows directly, the solution u of (1), $f \geqq 0$, $f \neq 0$, will not only take on its maximum for the same indices as f does but even conserve all order relations between components of f: *If* $f_i \geqq f_j$ *for some* i *and* j, *then* $u_i \geqq u_j$.

By induction and using necessary conditions analogical to that in Remark 4 to Def. 1, it can be shown that *any matrix conserving all order relations must be of the form* A_n.

Surely, any senseful maximum principle for matrices should be valid for A_n.

b) For $\Delta := a^3 + b^3 + 3abc - c^3 > 0$, consider the matrix

$$A = \begin{pmatrix} a & -c & b \\ b & a & -c \\ -c & b & a \end{pmatrix}.$$

Then

$$A^{-1} = \frac{1}{\Delta}\begin{pmatrix} a^2 + bc & b^2 + ac & c^2 - ab \\ c^2 - ab & a^2 + bc & b^2 + ac \\ b^2 + ac & c^2 - ab & a^2 + bc \end{pmatrix}.$$

Hence $A^{-1} > 0$ if, e.g., $0 < a < c$, $\max\,(a, c - a) < b < \dfrac{c^2}{a}$.

Inspite of Remark 1 to Theorem 1, that theorem is not applicable because (2) is violated. The weaker condition $Ae \geqq 0$ holds, but the maximum principle is not true:

Due to $b^2 + ac > a^2 + bc$ the main diagonal elements of A^{-1} are not dominating the other elements of their column, see now Remark 4 to Def. 1.

c) Let be P a permutation matrix. Clearly, P is of monotone kind (P^{-1} being a permutation matrix, too) but the maximum principle holds only if $P = I$. Now let be $P \neq I$. It can be shown that $P + \varepsilon I$ is of monotone kind if and only if $\varepsilon \in (-1, 0]$, condition (2) is true, but none of these matrices satisfies the maximum principle.

For the exceptional role of permutation matrices in another context, see [16] (or [17], p. 16).

d) We consider the weighted difference scheme (see [1], p. 278) for the constant coefficient one-dimensional heat equation,

$$y_t = y^{(\sigma)}_{\bar{x}x}\,, \qquad 0 < x_i < 1\,, \qquad t_j \geqq 0\,, \tag{12}$$

where

$$y_t := (y^{j+1}(x_i) - y^j(x_i))/\tau\,, \qquad y_{\bar{x}} := (y^j(x_i) - y^j(x_{i-1}))/h\,,$$
$$y_x := (y^j(x_{i+1}) - y^j(x_i))/h\,, \qquad y^{(\sigma)} := \sigma y^{j+1}(x_i) + (1 - \sigma)\, y^j(x_i)\,,$$
$$0 \leqq \sigma \leqq 1\,, \qquad t_j := j\tau\,, \qquad j = 0, 1, \dots\,, \qquad x_i := ih\,, \qquad i = 0, \dots, N\,,$$

that is, τ and h are the time and spatial discretization steps, respectively, and (different from Point 2) N is now an integer which we assume to be not less 2.

Along with (12) let be given initial and first kind boundary conditions. Introducing

$$\lambda := \tau/h^2\,,$$

the matrices $B_j = B$ and $C_j = C$ from Theorem 2 obey the form

$$B = \begin{pmatrix} 1 & 0 & 0 & \cdots & 0 \\ -\sigma\lambda & 1 + 2\sigma\lambda & -\sigma\lambda & \ddots & \vdots \\ 0 & \ddots & \ddots & \ddots & 0 \\ \vdots & \ddots & -\sigma\lambda & 1 + 2\sigma\lambda & -\sigma\lambda \\ 0 & \cdots & 0 & 0 & 1 \end{pmatrix}, \quad C = \begin{pmatrix} 1 & 0 & 0 & \cdots & 0 \\ (1-\sigma)\lambda & 1 - 2(1-\sigma)\lambda & (1-\sigma)\lambda & \ddots & \vdots \\ 0 & \ddots & \ddots & \ddots & 0 \\ \vdots & \ddots & (1-\sigma)\lambda & 1 - 2(1-\sigma)\lambda & (1-\sigma)\lambda \\ 0 & \cdots & 0 & 0 & 1 \end{pmatrix}.$$

B is seen to be an M matrix (since always $\sigma\lambda \geqq 0$), and C is nonnegative if

$$\sigma = 1 \quad \text{or} \quad \lambda \leqq 1/2(1 - \sigma)\,, \qquad 0 \leqq \sigma < 1\,. \tag{13}$$

Under these (well-known) conditions, $T_j = T$ in (9) is an M-matrix. Moreover, we have

$$T\begin{pmatrix} e \\ e \end{pmatrix} = \begin{pmatrix} e \\ 0 \end{pmatrix} \geqq 0\,.$$

Then, by Theorem 1, the maximum principle holds for (12).

Suppose now $N \geqq 3$, $\sigma < 1$ and $\lambda > 1/2(1-\sigma)$. We are going to apply Theorem 3 in [12]. For this aim, we partition T into the diagonal part T_d, the positive off-diagonal part $T^{(+)}$ and two negative off-diagonal parts T^z and T^s (comp. [12]),

$$T^{(+)} := \begin{pmatrix} 0 & 0 \\ -C^- & 0 \end{pmatrix}, \qquad T^z := \begin{pmatrix} 0 & 0 \\ 0 & B^- \end{pmatrix}, \qquad T^s := \begin{pmatrix} 0 & 0 \\ -C^+ & 0 \end{pmatrix}.$$

All conditions of [12] are found to be satisfied if

$$2(1-\sigma)\lambda - 1 \leqq (-\sigma\lambda)\frac{1}{1+2\sigma\lambda}(-(1-\sigma)\lambda), \tag{14}$$

which is, explicitely,

$$\lambda \leqq \frac{1}{3\sigma(1-\sigma)}\left(\sqrt{1-\sigma(1-\sigma)} - 1 + 2\sigma\right). \tag{15}$$

(14) corresponds to (4) in [12]. For $\sigma = \frac{1}{2}$, (15) amounts to $\lambda \leqq \frac{2}{3}\sqrt{3}$.

Performing a similar analysis for $N \geqq 2$ and second kind boundary conditions (approximated in the usual way [1], p. 296), instead of (15) the inequality

$$\lambda \leqq \frac{1}{2\sigma(1-\sigma)}\left(\sqrt{1-2\sigma(1-\sigma)} - 1 + 2\sigma\right) \tag{16}$$

is found. For $\sigma = \frac{1}{2}$ this means $\lambda \leqq \sqrt{2}$.

Under the conditions (15) or (16), T is by [12] of monotone kind, and, since $T\begin{pmatrix} e \\ e \end{pmatrix} \geqq 0$ remains true, the maximum principle [10] holds. (T is just of the structure considered in [10]. The above Theorem 1 does not apply here: (2) is violated.)

e) Next we investigate the problem of conservation of monotonicity for the scheme (12), with initial and first kind (constant) boundary values prescribed.

All conditions of Theorem 2 turn out to be met if (13) holds: B is an M-matrix (see above), $D = I$, $Ae = 0$, and the inequalities $c_{ii-1} \geqq 0$, $c_{i-1i} \geqq 0$, $c_{ii-1} + c_{i-1i} \leqq 1$ are equivalent to (13).

Using [12] in the case $N \geqq 4$, $\lambda > 1/2(1-\sigma)$, we find, similarly as above, T^S to be of monotone kind if (16) is true. (For comparison we remark that on the infinite x-line and for $\sigma = \frac{1}{2}$ in [15] the sharp bound $\lambda \leqq \frac{3}{2}$ has been established by calculating the exact solution y^1 to a step function y^0.)

Analogous considerations for second kind boundary conditions and $N \geqq 2$ lead to a condition stronger than (13):

$$\lambda \leqq \frac{1}{16\sigma(1-\sigma)}\left(\sqrt{9-4\sigma(1-\sigma)} - 3 + 6\sigma\right), \tag{17}$$

i.e., $\lambda \leqq \frac{1}{2}\sqrt{2}$ for $\sigma = \frac{1}{2}$.

Since for $N = 2$ and λ exceeding the bound (17) the corresponding matrix $(T^S)^{-1}$ has, indeed, negative elements, (12) does not conserve monotonicity in that case (see Remark 2 to Lemma 1). At the same time, as long as (13) or (16) are true, the maximum principles by Def. 1 or [10], respectively, are satisfied.

f) Finally, we consider the weighted difference scheme for the heat equation with variable coefficient $a(x) > 0$,

$$y_t = \frac{1}{h}(a_{i+1/2}y_x - a_{i-1/2}y_{\bar{x}})^{(\sigma)}, \tag{18}$$

along with initial and first kind boundary conditions.

Theorem 1 assures the maximum principle to hold if

$$1 \geqq (1-\sigma)\lambda(a_{i+1/2} + a_{i-1/2}), \qquad i = 1, \dots, N-1 \tag{19}$$

By Theorem 2, monotonicity will be conserved if

$$1 \geqq (1-\sigma)\lambda a_{i-1/2}, \qquad i = 1, N, \qquad 1 \geqq 2(1-\sigma)\lambda a_{i+1/2}, \qquad i = 1, \dots, N-2. \tag{20}$$

Obviously, for sufficiently small $(1-\sigma)\lambda$ both (19) and (20) are satisfied. For example, if

$$\tfrac{1}{2} \geqq (1-\sigma)\lambda a_{i-1/2}, \qquad i = 1, \dots, N. \tag{21}$$

But clearly, (19) may be true and, simultaneously, (20) not, and vice versa. Since both conditions are sufficient ones for the maximum principle and conservation of monotonicity to hold, respectively, we give explicit examples which show these properties to be, in fact, independent for (18):

1. $N = 3$, $\sigma = \frac{1}{2}$, $\lambda a_{1/2} = \lambda a_{5/2} = \frac{1}{3}$, $\lambda a_{3/2} = \frac{4}{3}$, $y^j(0) = 0$, $y^j(1) = 1$, $y^0 = (0, 0, 1, 1)^T$.

The exact solution y^1 of (18) for these data is readily calculated to be $y^1 = (0, 8/15, 7/15, 1)^T$, that is, the monotonicity of y^0 is not conserved. The maximum principle holds: (19) is true with strict inequality.

2. $N = 3$, $\sigma = \frac{1}{2}$, $\lambda a_{1/2} = \lambda a_{5/2} = \frac{8}{5}$, $\lambda a_{3/2} = \frac{3}{5}$

The conditions (20) are seen to hold with strict inequality, but (18) turns out to be not of monotone kind: The inverse of the matrix T_1 (see (9)) contains two negative elements. If, for example, $y^0 = (0, 1, 0, 0)^T$ and $y^1(0) = y^1(1) = 0$, then $y^1 = (0, -1/36, 5/36, 0)^T$, and the maximum principle is violated.

5. Completely monotone two-level methods

If the original initial-boundary value problem conserves monotonicity and satisfies the maximum principle, then it is natural to require that the corresponding discretized problem has these properties, too.

In view of the above examples, even well-known methods have to be reinvestigated whether they meet this requirement for which it may be useful to introduce a special term.

Definition 3. A two-level method (5) is called *completely monotone* if it satisfies the maximum principle and conserves monotonicity.

By Point 4f, we have already the following result.

Theorem 3: *For first kind boundary conditions the difference scheme* (18) *is completely monotone if* (21) *holds.*

Remark: For second kind boundary conditions, it suffices to require, additionally,

$$\tfrac{1}{3} \geqq (1 - \sigma)\, \lambda a_{i-1/2}\,, \qquad i = 1, N\,.$$

Thanks are due to M. DRYJA, Warsaw, for a copy of the paper [10].

References

1 Самарский, А. А., Теория разностных схем, Наука, Москва 1977.
2 STRANG, G.; FIX, G. J., An Analysis of the Finite Element Method, Prentice-Hall, Englewood Cliffs 1973.
3 COLLATZ, L., Aufgaben monotoner Art, Arch. der Math. 3 (1952), 365—376.
4 COLLATZ, L., Funktionalanalysis und Numerische Mathematik, Springer, Berlin 1964.
5 SCHRÖDER, J., Lineare Operatoren mit positiver Inversen, Arch. Rat. Mech. Analysis 8 (1961), 408—434.
6 VARGA, R. S., Matrix Iterative Analysis, Prentice-Hall, Englewood Cliffs 1962.
7 ORTEGA, J. M.; RHEINBOLDT, W. C., Iterative Solution of Nonlinear Equations in Several Variables, Acad. Press, N.Y. 1970.
8 BOHL, E., Monotonie: Lösbarkeit und Numerik bei Operatorgleichungen, Springer, Berlin 1974.
9 VARGA, R. S., On a discrete maximum principle, SIAM J. Numer. Analysis 3 (1966), 355—359.
10 CIARLET, P. G., Discrete maximum principle for finite-difference operators, Aequationes math. 4 (1970), 338—352.
11 STIELTJES, T. J., Sur les racines de l'équation $X_n = 0$, Acta Math. 9 (1887), 385—400.
12 LORENZ, J., Zur Inversmonotonie diskreter Probleme, Numer. Math. 27 (1977), 227—238.
13 Яненко, Н. Н., Об одном разностном методе счета многомерного уравнения теплопроводности. ДАН СССР 125, 6 (1959), 1207—1210.
14 Самарский, А. А., Об одном экономичном разностном методе решения многомерного параболического уравнения в произвольной области, Ж. вычисл. матем. и матем. физ. 2, 5 (1962), 787—811.
15 Годунов, С. К., Разностный метод численного расчета разрывных решений уравнений гидродинамики, Матем. сб. 47, 3 (1959), 271—306.
16 BIRKHOFF, G., Three observations on linear algebra. Universidad Nacional de Tucuman, Revista, Ser. A, 5 (1946), 147—151.
17 SCHAEFER, H. H., Banach Lattices and Positive Operators, Springer, Berlin 1974.

Eingereicht: 24. 8. 1981

Anschrift: Dr. GISBERT STOYAN, Inst. f. Math. d. AdW d. DDR, Mohrenstraße 39, DDR-1080 Berlin

ZAMM 62, 383 – 390 (1982)

V. I. Fabrikant / T. S. Sankar / L. M. Roytman / M. N. S. Swamy

Closed Form Solution to the Electrostatic Potential Problem for a Spherical Cap

Die folgende Aufgabe zum elektrostatischen Potential wird betrachtet: Gegeben ist eine beliebige Potentialverteilung auf einer Kugelkalotte, dann bestimme man die Verteilung der elektrischen Ladung. Man erhält eine Lösung in geschlossener Form durch eine spezielle Integraldarstellung des Kerns der kontrollierenden Integralgleichung. Es werden zwei Beipiele betrachtet: homogene Potentialverteilung über einer Kugelkalotte und der Fall einer geerdeten leitenden Kugelkalotte in einem homogenen elektrischen Feld. Die Ausdrücke für die Verteilung der elektrischen Ladung und die Gesamtladung werden angegeben. Die allgemeinen Ergebnisse dieser Arbeit können auch auf die verwandten Probleme einer Flüssigkeitsströmung und der Wärmeleitfähigkeit angewendet werden.

The following electrostatic potential problem is considered: arbitrary potential distribution is given on a spherical cap, the electric charge distribution is to be determined. The closed form solution is obtained by a special integral representation of the kernel of governing integral equation. Two examples are considered: uniform potential distribution over a spherical cap, and the case of earthed conducting spherical cap in a uniform electrical field. The expressions for the electric charge distribution and the total charge are given. The general results of this paper may be also applied to the related problems of a fluid flow and thermoconductivity.

Рассматривается следующая задача теории электростатического потенциала: дано произвольное распределение потенциала на поверхности сферического сегмента; требуется определить плотность распределения зарядов. Точное решение в замкнутой форме получено с помощью специального интегрального представления ядра основного интегрального уравнения. Рассмотрены два примера: равномерное распределение потенциала на поверхности сферического сегмента и случай заземлённого сферического сегмента в однородном электрическом поле. Даны выражения для плотности распределения заряда, а также определён полный заряд на поверхности. Общие результаты данной статьи могут быть использованы при решении математически аналогичных задач гидродинамики и теплопроводности.

1. Introduction

The solutions to potential problems have many applications in various branches of natural sciences. Despite the wide use of computers, exact solutions in closed form are still important, because they allow on the one hand to appreciate the accuracy of certain numerical method, and on the other hand, exact solution to a problem allows to formulate and to give an approximate solution to a more complicated problem.

The axisymmetric electrostatic potential problem for a spherical cap was solved by Collins [1]. He introduced a new unknown function $g(\theta)$ as the density of a special integral representation of the potential of a spherical cap, without any discussion of the physical meaning of this function. This approach allowed him to find the potential of the spherical cap electric field, but he did not discuss the definition of the electric charge density.

Asymmetric problem was solved recently in series by Homentcovschi [2]. He used expansion of the potential and the electric charge density as the Fourier series due to the spherical coordinate φ and established exact relationships between relevant harmonics in terms of Abel operators.

In this paper exact solution in closed form is given for the generalized asymmetric potential problem. The problem is called *generalized* due to the assumption that the potential of a point charge diminishes with distance R as $R^{1+\nu}$, $|\nu| < 1$. Electrostatic potential problem is a particular case of the generalized one, provided $\nu = 0$. The mathematical formulation of the problem is given in Section 2. The exact solution to the problem is presented in Section 3. In Section 4 some examples are presented, and the results obtained are discussed in Section 5.

2. Formulation of the Problem

Consider a spherical cap whose surface in spherical coordinates is described as $r = a$, $\theta \leqq \alpha$, $0 \leqq \varphi < 2\pi$. Arbitrary potential $u(\theta, \varphi)$ is prescribed at each point of the cap. The charge density $q(\theta, \varphi)$ is to be defined.

Assume the following generalized relationship between a point charge q, potential u and the distance from the charge R

$$u = qR^{-1-\nu} \quad \text{and} \quad -1 < \nu < 1 \,. \tag{1}$$

It is obvious that electrostatic potential problem is a particular case of the generalized one, when $\nu = 0$. Due to the assumption (1), *potential of the field of a charged spherical cap at arbitrary point of space* with spherical coordinates (r, θ, φ) will be defined as

$$U(r, \theta, \varphi) = \int_0^\alpha \int_0^{2\pi} \frac{q(\theta_0, \varphi_0)\, a^2 \sin\theta_0 \, d\theta_0 \, d\varphi_0}{\{r^2 + a^2 - 2ar[\cos\theta\cos\theta_0 + \sin\theta\sin\theta_0\cos(\varphi - \varphi_0)]\}^{(1+\nu)/2}} \,. \tag{2}$$

As the potential at the surface of the cap is given, expression (2) leads to the following integral equation with respect to unknown charge density

$$\frac{a^{1-\nu}}{2^{(1+\nu)/2}} \int_0^\alpha \int_0^{2\pi} \frac{q(\theta_0, \varphi_0) \sin\theta_0 \, d\theta_0 \, d\varphi_0}{\{1 - [\cos\theta\cos\theta_0 + \sin\theta\sin\theta_0\cos(\varphi - \varphi_0)]\}^{(1+\nu)/2}} = U(a, \theta, \varphi) = u(\theta, \varphi) \,. \tag{3}$$

Now transform the denominator in the integral (3)

$$1 - [\cos\theta\cos\theta_0 + \sin\theta\sin\theta_0\cos(\varphi-\varphi_0)] = -[\tfrac{1}{2}\sin\theta\sin\theta_0\,(e^{i(\varphi-\varphi_0)} + e^{-i(\varphi-\varphi_0)}) + \cos\theta\cos\theta_0 - 1]$$
$$= -\frac{1}{2}\sin\theta\sin\theta_0\, e^{-i(\varphi-\varphi_0)}\left[e^{2i(\varphi-\varphi_0)} + 2\frac{\cos\theta\cos\theta_0 - 1}{\sin\theta\sin\theta_0}e^{i(\varphi-\varphi_0)} + 1\right]. \quad (4)$$

Use the identity

$$2\frac{\cos\theta\cos\theta_0 - 1}{\sin\theta\sin\theta_0} = \frac{\cos\theta\cos\theta_0 + \cos\theta - \cos\theta_0 - 1 + \cos\theta\cos\theta_0 - \cos\theta + \cos\theta_0 - 1}{\sin\theta\sin\theta_0}$$
$$= \frac{(\cos\theta_0 + 1)(\cos\theta - 1) + (\cos\theta_0 - 1)(\cos\theta + 1)}{\sin\theta\sin\theta_0} = -\left(\frac{\operatorname{tg}\frac{\theta}{2}}{\operatorname{tg}\frac{\theta_0}{2}} + \frac{\operatorname{tg}\frac{\theta_0}{2}}{\operatorname{tg}\frac{\theta}{2}}\right). \quad (5)$$

Substitution of (5) into (4) leads to

$$1 - [\cos\theta\cos\theta_0 + \sin\theta\sin\theta_0\cos(\varphi-\varphi_0)] =$$
$$= -\frac{1}{2}\sin\theta\sin\theta_0\, e^{-i(\varphi-\varphi_0)}\left(e^{i(\varphi-\varphi_0)} - \frac{\operatorname{tg}\frac{\theta}{2}}{\operatorname{tg}\frac{\theta_0}{2}}\right)\left(e^{i(\varphi-\varphi_0)} - \frac{\operatorname{tg}\frac{\theta_0}{2}}{\operatorname{tg}\frac{\theta}{2}}\right)$$
$$= \frac{\sin\theta\sin\theta_0}{2}\cdot\frac{\operatorname{tg}\frac{\theta_0}{2}}{\operatorname{tg}\frac{\theta}{2}}\left(1 - e^{-i(\varphi-\varphi_0)}\frac{\operatorname{tg}\frac{\theta}{2}}{\operatorname{tg}\frac{\theta_0}{2}}\right)\left(1 - e^{i(\varphi-\varphi_0)}\frac{\operatorname{tg}\frac{\theta}{2}}{\operatorname{tg}\frac{\theta_0}{2}}\right).$$
$$= \frac{\sin\theta\sin\theta_0}{2\operatorname{tg}\frac{\theta}{2}\operatorname{tg}\frac{\theta_0}{2}}\left[\operatorname{tg}^2\frac{\theta}{2} + \operatorname{tg}^2\frac{\theta_0}{2} - 2\operatorname{tg}\frac{\theta}{2}\operatorname{tg}\frac{\theta_0}{2}\cos(\varphi-\varphi_0)\right].$$

And the final simplification gives

$$1 - [\cos\theta\cos\theta_0 + \sin\theta\sin\theta_0\cos(\varphi-\varphi_0)] = 2\cos^2\frac{\theta}{2}\cos^2\frac{\theta_0}{2}\left[\operatorname{tg}^2\frac{\theta}{2} + \operatorname{tg}^2\frac{\theta_0}{2} - \right.$$
$$\left. - 2\operatorname{tg}\frac{\theta}{2}\operatorname{tg}\frac{\theta_0}{2}\cos(\varphi-\varphi_0)\right]. \quad (6)$$

Substitution of (6) into (3) yields the governing integral equation in the form

$$\frac{a^{1-\nu}}{2^{1+\nu}}\int_0^\alpha\int_0^{2\pi}\frac{q(\theta_0,\varphi_0)\sin\theta_0\,d\theta_0\,d\varphi_0}{\left(\cos\frac{\theta}{2}\cos\frac{\theta_0}{2}\right)^{1+\nu}\left[\operatorname{tg}^2\frac{\theta}{2} + \operatorname{tg}^2\frac{\theta_0}{2} - 2\operatorname{tg}\frac{\theta}{2}\operatorname{tg}\frac{\theta_0}{2}\cos(\varphi-\varphi_0)\right]^{(1+\nu)/2}} = u(\theta,\varphi)$$
$$\text{for}\quad 0 \leqq \theta \leqq \alpha, \qquad 0 \leqq \varphi < 2\pi. \quad (7)$$

3. Solution

Here the exact solution of (7) in closed form is given by mean of a special integral representation of the kernel of equation (7). Introduce notations

$$z = \operatorname{tg}\frac{\theta}{2}, \qquad z_0 = \operatorname{tg}\frac{\theta_0}{2}. \quad (8)$$

Transform the denominator of (7)

$$\frac{1}{[z^2 + z_0^2 - 2zz_0\cos(\varphi-\varphi_0)]^{(1+\nu)/2}}$$
$$= \sum_{n=-\infty}^{\infty}\frac{e^{in(\varphi-\varphi_0)}}{2\pi}\int_0^{2\pi}\frac{e^{-in\psi}\,d\psi}{[z^2 + z_0^2 - 2zz_0\cos\psi]^{(1+\nu)/2}}$$
$$= \sum_{n=-\infty}^{\infty}\frac{e^{in(\varphi-\varphi_0)}}{2\pi z_0^{1+\nu}}\,\frac{2\pi\Gamma[|n| + (1+\nu)/2]}{\Gamma[(1+\nu)/2]\,\Gamma(|n|+1)}\left(\frac{z}{z_0}\right)^{|n|}{}_2F_1\left(\frac{1+\nu}{2}, |n| + \frac{1+\nu}{2}, |n|+1; \frac{z^2}{z_0^2}\right)$$

$$= \sum_{n=-\infty}^{\infty} \frac{e^{in(\varphi-\varphi_0)}}{z_0^{1+\nu}} \cdot \frac{\Gamma[|n| + (1+\nu)/2]}{\Gamma[(1+\nu/2]\,\Gamma(|n|+1)} \left(\frac{z}{z_0}\right)^{|n|} \times$$

$$\times \frac{2\Gamma(|n|+1)}{\Gamma[|n| + (1+\nu)/2]\,\Gamma[1-(1+\nu)/2]} \int_0^1 \frac{t^{2|n|+\nu}(1-t^2)^{-(1+\nu)/2}}{[(1-t^2z^2/z_0^2)]^{(1+\nu)/2}}\,dt$$

$$= \sum_{n=-\infty}^{\infty} \frac{2}{\pi} \cos\frac{\pi\nu}{2} \frac{e^{in(\varphi-\varphi_0)}}{(zz_0)^{|n|}} \int_0^z \frac{x^{2|n|+\nu}\,dx}{[(z^2-x^2)\,(z_0^2-x^2)]^{(1+\nu)/2}}. \tag{9}$$

Here the Fourier series expansion was used, the integral 2.4.10 and integral representation of hypergeometric function 2.12.1 from [3] were employed. All the transformations (9) are valid for $z < z_0$, when $z > z_0$ everything remains the same, except the upper limit of integration in (9) should be changed to z_0. Summation of (9) finally gives the necessary integral representation

$$\frac{1}{[z^2 + z_0^2 - 2zz_0 \cos(\varphi - \varphi_0)]^{(1+\nu)/2}}$$

$$= \frac{2}{\pi} \cos\frac{\pi\nu}{2} \int_0^{\min(z,z_0)} \frac{x^\nu\,dx}{[(z^2-x^2)\,(z_0^2-x^2)]^{(1+\nu)/2}} \left(\frac{1}{1 - x^2 e^{i(\varphi-\varphi_0)}/zz_0} + \frac{1}{1 - x^2 e^{-i(\varphi-\varphi_0)}/zz_0} - 1\right). \tag{10}$$

Having in mind notation (8), it is convenient to denote $x = \mathrm{tg}\,(\xi/2)$, $w = \mathrm{tg}^2 \frac{\xi}{2} \Big/ \mathrm{tg}\frac{\theta}{2}\,\mathrm{tg}\frac{\theta_0}{2}$, and (10) transforms to

$$\frac{1}{\left[\mathrm{tg}^2\frac{\theta}{2} + \mathrm{tg}^2\frac{\theta_0}{2} - 2\,\mathrm{tg}\frac{\theta}{2}\,\mathrm{tg}\frac{\theta_0}{2}\cos(\varphi-\varphi_0)\right]^{(1+\nu)/2}}$$

$$= \frac{2}{\pi}\cos\frac{\pi\nu}{2} \int_0^{\min(\theta,\theta_0)} \frac{\mathrm{tg}^\nu\frac{\xi}{2}\,d\xi}{2\cos^2\frac{\xi}{2}\left[\left(\mathrm{tg}^2\frac{\theta}{2} - \mathrm{tg}^2\frac{\xi}{2}\right)\left(\mathrm{tg}^2\frac{\theta_0}{2} - \mathrm{tg}^2\frac{\xi}{2}\right)\right]^{(1+\nu)/2}} \times$$

$$\times \left(\frac{1}{1 - w\,e^{i(\varphi-\varphi_0)}} + \frac{1}{1 - w\,e^{-i(\varphi-\varphi_0)}} - 1\right). \tag{11}$$

Use the identity

$$\mathrm{tg}^2\frac{\theta}{2} - \mathrm{tg}^2\frac{\xi}{2} = \frac{2}{1+\cos\theta} - \frac{2}{1+\cos\xi} = \frac{\cos\xi - \cos\theta}{2\cos^2\frac{\theta}{2}\cos^2\frac{\xi}{2}}. \tag{12}$$

Substitution of (12) into (11) gives

$$\frac{1}{\left[\mathrm{tg}^2\frac{\theta}{2} + \mathrm{tg}^2\frac{\theta_0}{2} - 2\,\mathrm{tg}\frac{\theta}{2}\,\mathrm{tg}\frac{\theta_0}{2}\cos(\varphi-\varphi_0)\right]^{(1+\nu)/2}} =$$

$$= \frac{2}{\pi}\cos\frac{\pi\nu}{2}\left(\cos\frac{\theta}{2}\cos\frac{\theta_0}{2}\right)^{1+\nu} \int_0^{\min(\theta,\theta_0)} \frac{\sin^\nu\xi\,d\xi}{[(\cos\xi - \cos\theta)(\cos\xi - \cos\theta_0)]^{(1+\nu)/2}} \times$$

$$\times \left(\frac{1}{1 - w\,e^{i(\varphi-\varphi_0)}} + \frac{1}{1 - w\,e^{-i(\varphi-\varphi_0)}} - 1\right). \tag{13}$$

Introduce L-operator as

$$L(w, \varphi - \varphi_0)\,f(\varphi_0) = \int_0^{2\pi} \left(\frac{1}{1 - w\,e^{i(\varphi-\varphi_0)}} + \frac{1}{1 - w\,e^{-i(\varphi-\varphi_0)}} - 1\right) f(\varphi_0)\,d\varphi_0. \tag{14}$$

Substitution of (13) into (7), taking into consideration (14) yields after changing of the order of integration

$$\frac{a^{1-\nu}\cos\frac{\pi\nu}{2}}{2^\nu\pi} \int_0^\theta \frac{\sin^\nu\xi\,d\xi}{(\cos\xi - \cos\theta)^{(1+\nu)/2}} \int_\xi^\alpha \frac{\sin\theta_0\,d\theta_0}{(\cos\xi - \cos\theta_0)^{(1+\nu)/2}} L\left(\frac{\mathrm{tg}^2\frac{\xi}{2}}{\mathrm{tg}\frac{\theta}{2}\,\mathrm{tg}\frac{\theta_0}{2}}, (\varphi-\varphi_0)\right) q(\theta_0, \varphi_0)$$

$$= u(\theta, \varphi). \tag{15}$$

Here the following scheme of changing of the order of integration was employed

$$\int_0^\theta d\theta_0 \int_0^{\theta_0} d\xi + \int_\theta^\alpha d\theta_0 \int_0^\theta d\xi = \int_0^\theta d\xi \int_\xi^\theta d\theta_0 + \int_0^\theta d\xi \int_\theta^\alpha d\theta_0 = \int_0^\theta d\xi \int_\xi^\alpha d\theta_0 . \tag{16}$$

Now integral equation (15) is represented as sequence of two operators of Abel type and L-operator. The inverse Abel operators are known, and it remains to find the inverse L-operator. The following properties of L-operators can be easily verified

$$L(w_1, \varphi_1 - \varphi)\, L(w, \varphi - \varphi_0)\, f(\varphi_0) = 2\pi L(ww_1, \varphi_1 - \varphi_0)\, f(\varphi_0)\,, \qquad L(1, \varphi - \varphi_0)\, f(\varphi_0) = 2\pi f(\varphi)\,. \quad (17), (18)$$

These properties allow to construct the inverse operator

$$L^{-1}(w, \varphi - \varphi_0) = \frac{1}{2\pi} L\left(\frac{1}{w}, \varphi - \varphi_0\right). \tag{19}$$

Hereafter make use of the integral

$$\int_{\theta_1}^{0} \frac{\sin \xi \, d\xi}{(\cos \xi - \cos \theta)^{(1+\nu)/2} (\cos \theta_1 - \cos \xi)^{(1-\nu)/2}} = \frac{\Gamma\left(\frac{1-\nu}{2}\right) \Gamma\left(\frac{1+\nu}{2}\right)}{\Gamma(1)} = \frac{\pi}{\cos \frac{\pi\nu}{2}} . \tag{20}$$

Apply to both parts of (15) the operator

$$\frac{d}{d\theta_1} \int_0^{\theta_1} \frac{\sin \theta_1 \, d\theta_1}{(\cos \theta - \cos \theta_1)^{(1-\nu)/2}} L\left(\operatorname{tg} \frac{\theta}{2}, \varphi_1 - \varphi\right).$$

Using (17) and (20), the result is

$$(2a)^{1-\nu} \pi \sin^\nu \theta_1 \int_{\theta_1}^{\alpha} \frac{\sin\theta_0 \, d\theta_0}{(\cos \theta_1 - \cos \theta_0)^{(1+\nu)/2}} L\left(\frac{\operatorname{tg} \frac{\theta_1}{2}}{\operatorname{tg} \frac{\theta_0}{2}}, \varphi_1 - \varphi_0\right) q(\theta_0, \varphi_0)$$

$$= \frac{d}{d\theta_1} \int_0^{\theta_1} \frac{\sin \theta_1 \, d\theta_1}{(\cos \theta - \cos \theta_1)^{(1-\nu)/2}} L\left(\operatorname{tg} \frac{\theta}{2}, \varphi_1 - \varphi\right) u(\theta, \varphi) = \chi(\theta_1, \varphi_1)\,. \tag{21}$$

Application of the operator

$$\frac{d}{d\theta_2} \int_{\theta_2}^{\alpha} \frac{\sin^{1-\nu} \theta_1 \, d\theta_1}{(\cos \theta_2 - \cos \theta_1)^{(1-\nu)/2}} L\left(\frac{1}{\operatorname{tg}^2 \frac{\theta_1}{2}}, \varphi_2 - \varphi_1\right)$$

to both parts of (21) yields

$$-\frac{(2a)^{1-\nu}\, 2\pi^3}{\cos \frac{\pi\nu}{2}} \sin \theta_2\, L\left(\frac{1}{\operatorname{tg} \frac{\theta_2}{2}}, \varphi_2 - \varphi_0\right) q(\theta_2, \varphi_0)$$

$$= \frac{d}{d\theta_2} \int_{\theta_2}^{\alpha} \frac{\sin^{1-\nu} \theta_1 \, d\theta_1}{(\cos \theta_2 - \cos \theta_1)^{(1-\nu)/2}} L\left(\frac{1}{\operatorname{tg}^2 \frac{\theta_1}{2}}, \varphi_2 - \varphi_1\right) \chi(\theta_1, \varphi_1)\,. \tag{22}$$

And finally application of the operator $L\left(\operatorname{tg} \frac{\theta_2}{2}, \varphi_3 - \varphi_2\right)$ to both parts of (22), taking into consideration (18), (19) and (21), gives the closed form solution

$$q(\theta_2, \varphi_3) = -\frac{\cos (\pi\nu/2)}{(2a)^{1-\nu}\, 8\pi^5 \sin \theta_2} L\left(\operatorname{tg} \frac{\theta_2}{2}, \varphi_3 - \varphi_2\right) \frac{d}{d\theta_2} \int_{\theta_2}^{\alpha} \frac{\sin^{1-\nu} \theta_1 \, d\theta_1}{(\cos \theta_2 - \cos\theta_1)^{(1-\nu)/2}} \times$$

$$\times L\left(\frac{1}{\operatorname{tg}^2 \frac{\theta_1}{2}}, \varphi_2 - \varphi_1\right) \frac{d}{d\theta_1} \int_0^{\theta_1} \frac{\sin \theta \, d\theta}{(\cos \theta - \cos \theta_1)^{(1-\nu)/2}} L\left(\operatorname{tg} \frac{\theta}{2}, \varphi_1 - \varphi\right) u(\theta, \varphi)\,. \tag{23}$$

Some further simplifications of (23) are possible using the properties of L-operators (17) and the following rules of differentiation under the integral sign

$$\frac{\mathrm{d}}{\mathrm{d}\theta_1}\int_0^{\theta_1}\frac{f(\theta)\sin\theta\,\mathrm{d}\theta}{(\cos\theta-\cos\theta_1)^{(1-\nu)/2}}=\sin\theta_1\left[\frac{f(0)}{(1-\cos\theta_1)^{(1-\nu)/2}}+\int_0^{\theta_1}\frac{\mathrm{d}\theta}{(\cos\theta-\cos\theta_1)^{(1-\nu)/2}}\frac{\mathrm{d}f(\theta)}{\mathrm{d}\theta}\right], \tag{24}$$

$$\frac{\mathrm{d}}{\mathrm{d}\theta_2}\int_{\theta_2}^{\alpha}\frac{f(\theta_1)\sin\theta_1\,\mathrm{d}\theta_1}{(\cos\theta_2-\cos\theta_1)^{(1-\nu)/2}}=\sin\theta_2\left[-\frac{f(\alpha)}{(\cos\theta_2-\cos\alpha)^{(1-\nu)/2}}+\int_{\theta_2}^{\alpha}\frac{\mathrm{d}\theta_1}{(\cos\theta_2-\cos\theta_1)^{(1-\nu)/2}}\frac{\mathrm{d}f(\theta_1)}{\mathrm{d}\theta_1}\right]. \tag{25}$$

The simplified result is

$$q(\theta_2,\varphi_2)=\frac{\cos(\pi\nu/2)}{(2a)^{1-\nu}\,2\pi^3}\left\{\frac{\Phi(\alpha,\theta_2,\varphi_2)}{(\cos\theta_2-\cos\alpha)^{(1-\nu)/2}}-\int_{\theta_2}^{\alpha}\frac{\mathrm{d}\theta_1}{(\cos\theta_2-\cos\theta_1)^{(1-\nu)/2}}\frac{\mathrm{d}}{\mathrm{d}\theta_1}\Phi(\theta_1,\theta_2,\varphi_2)\right\},$$

$$\Phi(\theta_1,\theta_2,\varphi_2)=\sin^{1-\nu}\theta_1\int_0^{2\pi}\left\{\frac{u(0,\varphi)}{(1-\cos\theta_1)^{(1-\nu)/2}}+\int_0^{\theta_1}\frac{\mathrm{d}\theta}{(\cos\theta-\cos\theta_1)^{(1-\nu)/2}}\times\right.$$
$$\left.\times\frac{\mathrm{d}}{\mathrm{d}\theta}\left[\frac{1}{1-\dfrac{\mathrm{tg}\frac{\theta_2}{2}\,\mathrm{tg}\frac{\theta}{2}}{\mathrm{tg}^2\frac{\theta_1}{2}}\mathrm{e}^{i(\varphi_2-\varphi)}}+\frac{1}{1-\dfrac{\mathrm{tg}\frac{\theta_2}{2}\,\mathrm{tg}\frac{\theta}{2}}{\mathrm{tg}^2\frac{\theta_1}{2}}\mathrm{e}^{-i(\varphi_2-\varphi)}}-1\right]u(\theta,\varphi)\right\}\mathrm{d}\varphi\,. \tag{26}$$

Now the exact solution in closed form to the generalized potential problem is given by either (23) or (26). The case of electrostatic potential problem is readily obtained by setting $\nu=0$. Instead of (26) one will have

$$q(\theta_2,\varphi_2)=\frac{1}{4a\pi^3}\left\{\frac{\Phi(\alpha,\theta_2,\varphi_2)}{\sqrt{\cos\theta_2-\cos\alpha}}-\int_{\theta_2}^{\alpha}\frac{\mathrm{d}\theta_1}{\sqrt{\cos\theta_2-\cos\theta_1}}\frac{\mathrm{d}}{\mathrm{d}\theta_1}\Phi(\theta_1,\theta_2,\varphi_2)\right\},$$

$$\Phi(\theta_1,\theta_2,\varphi_2)=\sin\theta_1\int_0^{2\pi}\left\{\frac{u(0,\varphi)}{\sqrt{2}\sin(\theta_1/2)}+\int_0^{\theta_1}\frac{\mathrm{d}\theta}{\sqrt{\cos\theta-\cos\theta_1}}\times\right.$$
$$\left.\times\frac{\mathrm{d}}{\mathrm{d}\theta}\left[\frac{1}{1-\dfrac{\mathrm{tg}\frac{\theta_2}{2}\,\mathrm{tg}\frac{\theta}{2}}{\mathrm{tg}^2\frac{\theta_1}{2}}\mathrm{e}^{i(\varphi_2-\varphi)}}+\frac{1}{1-\dfrac{\mathrm{tg}\frac{\theta_2}{2}\,\mathrm{tg}\frac{\theta}{2}}{\mathrm{tg}^2\frac{\theta_1}{2}}\mathrm{e}^{-i(\varphi_2-\varphi)}}-1\right]u(\theta,\varphi)\right\}\mathrm{d}\varphi\,. \tag{27}$$

4. Examples

Consider a *spherical cap under constant potential.* For this case $u(\theta,\varphi)=u_0=\text{const}$, and use of (27) gives

$$\Phi(\theta_1,\theta_2,\varphi_2)=2\pi\sqrt{2}\,u_0\cos(\theta_1/2)\,,\qquad\frac{\mathrm{d}}{\mathrm{d}\theta_1}\Phi(\theta_1,\theta_2,\varphi_2)=-\pi\sqrt{2}\,u_0\sin\frac{\theta_1}{2},$$

$$q(\theta)=\frac{u_0}{2\pi^2a}\left[\frac{\sqrt{2}\cos(\alpha/2)}{\sqrt{\cos\theta-\cos\alpha}}+\tan^{-1}\sqrt{\frac{\cos\theta-\cos\alpha}{1+\cos\alpha}}\right]. \tag{28}$$

The following integral was employed

$$\int_{\theta_2}^{\alpha}\frac{\sin\frac{\theta_1}{2}\,\mathrm{d}\theta_1}{\sqrt{\cos\theta_2-\cos\theta_1}}=-\sqrt{2}\int_{\theta_2}^{\alpha}\frac{\mathrm{d}\left(\cos\frac{\theta_1}{2}\right)}{\sqrt{\cos^2\frac{\theta_2}{2}-\cos^2\frac{\theta_1}{2}}}=\sqrt{2}\cos^{-1}\left(\frac{\cos\frac{\alpha}{2}}{\cos\frac{\theta_2}{2}}\right)$$
$$=\sqrt{2}\tan^{-1}\sqrt{\frac{\cos\theta_2-\cos\alpha}{1+\cos\alpha}}. \tag{29}$$

Integration of (28) gives the *total charge*

$$Q = 2\pi a^2 \int_0^\alpha q(\theta) \sin\theta \, d\theta = \frac{u_0 a}{\pi} (\sin\alpha + \alpha) . \tag{30}$$

The following integrals were employed

$$\int_0^\alpha \frac{\sin\theta \, d\theta}{\sqrt{\cos\theta - \cos\alpha}} = 2\sqrt{1 - \cos\alpha} = 2\sqrt{2} \sin(\alpha/2) , \tag{31}$$

$$\int_0^\alpha \tan^{-1} \sqrt{\frac{\cos\theta - \cos\alpha}{1 + \cos\alpha}} \sin\theta \, d\theta$$

$$= \Big|_0^\alpha \sqrt{1 + \cos\alpha} \sqrt{\cos\theta - \cos\alpha} - \Big|_0^\alpha (1 + \cos\theta) \tan^{-1} \sqrt{\frac{\cos\theta - \cos\alpha}{1 + \cos\alpha}} = -\sin\alpha + \alpha . \tag{32}$$

The result (30) corresponds to the known one [1]. The *electrical capacity* of the cap is

$$C = \frac{Q}{u_0} = \frac{a}{\pi} (\sin\alpha + \alpha) . \tag{33}$$

As the second example consider the case of *earthed spherical cap in a uniform electrical field E parallel to the axis of the cap.* For this case

$$u(\theta, \varphi) = Ea \cos\theta .$$

Formulae (27) give

$$\Phi(\theta_1, \theta_2, \varphi_2) = 2\pi Ea \sin\theta_1 \left[\frac{1}{\sqrt{1 - \cos\theta_1}} - 2\sqrt{1 - \cos\theta_1} \right] = 2\pi\sqrt{2}\, Ea \cos\frac{3}{2}\theta_1 ,$$

$$\frac{d}{d\theta_1} \Phi(\theta_1, \theta_2, \varphi_2) = -3\pi\sqrt{2}\, Ea \sin\frac{3}{2}\theta_1 ,$$

$$q(\theta) = \frac{E}{2\pi^2} \left[\frac{\sqrt{2}\cos\frac{3}{2}\alpha}{\sqrt{\cos\theta - \cos\alpha}} + 3\sqrt{2}\cos\frac{\alpha}{2}\sqrt{\cos\theta - \cos\alpha} + 3\cos\theta \tan^{-1}\sqrt{\frac{\cos\theta - \cos\alpha}{1 + \cos\alpha}} \right. . \tag{34}$$

Here the following integral was used

$$\int_{\theta_2}^\alpha \frac{\sin\frac{3}{2}\theta_1 \, d\theta_1}{\sqrt{\cos\theta_2 - \cos\theta_1}} = 2\sqrt{2} \Big|_{\theta_2}^\alpha \cos\frac{\theta_1}{2} \sqrt{\cos^2\frac{\theta_2}{2} - \cos^2\frac{\theta_1}{2}} - \sqrt{2}\cos\theta_2 \Big|_{\theta_2}^\alpha \sin^{-1}\left(\frac{\cos\frac{\theta_1}{2}}{\cos\frac{\theta_2}{2}} \right)$$

$$= 2\cos\frac{\alpha}{2}\sqrt{\cos\theta_2 - \cos\alpha} + \sqrt{2}\cos\theta_2 \tan^{-1}\sqrt{\frac{\cos\theta - \cos\alpha}{1 + \cos\alpha}} . \tag{35}$$

The *total charge* for this case is

$$Q = 2\pi a^2 \int_0^\alpha q(\theta) \sin\theta \, d\theta = \frac{Ea^2}{\pi} \sin\alpha (1 + \cos\alpha) . \tag{36}$$

The following integral was employed

$$\int_0^\alpha \sin\theta \cos\theta \tan^{-1} \sqrt{\frac{\cos\theta - \cos\alpha}{1 + \cos\alpha}} \, d\theta$$

$$= \Big|_0^\alpha \frac{\sin^2\theta}{2} \tan^{-1} \sqrt{\frac{\cos\theta - \cos\alpha}{1 + \cos\alpha}} + \sqrt{1 + \cos\alpha} \Big|_0^\alpha \left[\frac{1 - \cos\alpha}{2} \sqrt{\cos\theta - \cos\alpha} - \frac{1}{6} (\cos\theta - \right.$$

$$\left. - \cos\alpha)^{3/2} \right] = \frac{1}{3} (1 - \cos\alpha) \sin\alpha . \tag{37}$$

All other integrals involved are elementary.

5. Discussion

It is of interest to establish a relationship between the results of this paper and the previously obtained ones. COLLINS [1] introduced the following integral representation of the potential at the cap surface for axisymmetric case

$$u(\theta) = \int_0^\theta \frac{g(\eta) \sec(\eta/2) \, d\eta}{\sqrt{2(\cos\eta - \cos\theta)}} . \tag{38}$$

It is possible to establish the relationship between introduced by COLLINS unknown function $g(\eta)$ and the electric charge density q. Comparison of (38) and (15), taking into consideration, that q does not depend on φ and $\nu = 0$, immediately gives the sought relationship

$$g(\eta) = 2\sqrt{2}\, a \cos\frac{\eta}{2} \int_\eta^\alpha \frac{q(\theta_0) \sin\theta_0 \, d\theta_0}{\sqrt{\cos\eta - \cos\theta_0}} . \tag{39}$$

Substitution of (28) and (34) into (39) gives exactly the same results as the ones obtained in [1]. Comparison of (39) with (21), taking into consideration (24) and (26) gives even more simple relationship between the functions g and Φ

$$g(\theta) = \frac{1}{\sqrt{2}\pi^2} \cos\frac{\theta}{2}\, \Phi(\theta) . \tag{40}$$

If the following expansions are valid

$$u(\theta, \varphi) = \sum_{n=-\infty}^{\infty} u_n(\theta)\, e^{in\varphi} , \qquad q(\theta, \varphi) = \sum_{n=-\infty}^{\infty} q_n(\theta)\, e^{in\varphi} , \tag{41}$$

then substitution of (41) into (23) gives the relationship between relevant harmonics

$$q_n(\theta_2) = -\frac{\cos(\pi\nu/2)\, \mathrm{tg}^n \frac{\theta_2}{2}}{(2a)^{1-\nu} \pi^2 \sin\theta_2} \frac{d}{d\theta_2} \int_{\theta_2}^{\alpha} \frac{\sin^{1-\nu}\theta_1 \, d\theta_1}{\mathrm{tg}^{2n}(\theta_1/2) (\cos\theta_2 - \cos\theta_1)^{(1-\nu)/2}} \frac{d}{d\theta_1} \int_0^{\theta_1} \frac{\mathrm{tg}^n(\theta/2) \sin\theta \, u_n(\theta)}{(\cos\theta - \cos\theta_1)^{(1-\nu)/2}} \, d\theta . \tag{42}$$

For the case of electrostatic potential problem $\nu = 0$, and expression (42) corresponds to the result obtained by HOMENTCOVSCHI [2].

A general expression can be established for the total charge in terms of given potential. In general case

$$Q = \int_0^\alpha \int_0^{2\pi} q(\theta_2, \varphi_3)\, a^2 \sin\theta_2 \, d\varphi_3 . \tag{43}$$

Substitution of (23) in (43) and using of the properties of L-operators (17) results in

$$Q = \frac{a^{1+\nu} \cos(\pi\nu/2)}{2^{1-\nu}\pi^2} \int_0^\alpha \frac{\sin^{1-\nu}\theta_1 \, d\theta_1}{(1 - \cos\theta_1)^{(1-\nu)/2}} \frac{d}{d\theta_1} \int_0^{\theta_1} \frac{\sin\theta \, d\theta}{(\cos\theta - \cos\theta_1)^{(1-\nu)/2}} \int_0^{2\pi} u(\theta, \varphi) \, d\varphi . \tag{44}$$

Further simplification occurs in the case of electrostatic potential problem, when $\nu = 0$

$$Q = \frac{a}{\sqrt{2}\,\pi^2} \int_0^\alpha \cos\frac{\theta_1}{2} \, d\theta_1 \frac{d}{d\theta_1} \int_0^{\theta_1} \frac{\sin\theta \, d\theta}{\sqrt{\cos\theta - \cos\theta_1}} \int_0^{2\pi} u(\theta, \varphi) \, d\varphi .$$

The last result corrects small misprints which occured in the paper [2] (formula 10).

All the results of this paper may also be used for the solution of several problems in hydromechanics and thermoconductivity which have similar mathematical formulation.

Acknowledgement

The authors acknowledge the support given by the National Research Council of Canada under grants A-7739 and A-7104, and also by Pennsylvania State University in carrying out this research.

References

1 Collins, W. D., On the solution of some axisymmetric boundary value problems by means of integral equations, Quart. J. Mech. Appl. Math. 12 (1959), 232–241
2 Homentcovschi, D., On the electrostatic potential problem for a spherical cap, ZAMM 60 (1980), 636–637.
3 Bateman, G.; Erdelyi, A., Higher Transcendental Functions. Vol. 1, McGraw-Hill Book Company Inc. 1953.

Eingereicht: 26. 8. 1981

Anschriften: Dr. V. I. Fabrikant, Dr. T. S. Sankar, Mechanical Engineering Department; Dr. L. M. Roytman, Dr. M. N. S. Swamy, Electrical Engineering Department, Concordia University, Sir George Williams Campus, 1455 De Maisonneuve Boulevard West, Montreal, Quebec H3G 1M8 Canada

ZAMM 62, 391 – 399 (1982)

W. C.-C. Yeung / M. Holt

Boundary Layer Control in Pipes through Strong Injection

Bei der Kohlevergasung erzeugen Oxidation und Sulfidbildung in Rohren schwerwiegende Korrosion. In der vorliegenden Arbeit wird versucht, festzustellen, ob eine Verminderung derartiger Korrosion möglich ist, wenn Dampf am Rohranfang eingespritzt wird mit dem Ziel, die Gaszusammensetzung in der Grenzschicht entlang der Rohrwände zu verändern. Die Dampfzuführung bildet eine dünne Schicht an der Innenwand, die für eine gewisse Zeit vor einem Kontakt mit dem korrosionserzeugenden Gas schützt. (Turbulenz wird unter Umständen die Diffusion durch die schützende Schicht beschleunigen.) Als Gasprodukte werden Schwefelwasserstoff und Wasserdampf angenommen. Um numerische Lösungen für die bestimmenden Gleichungen zu erhalten, wird die Methode der Integralrelationen verwendet. Die Konzentration von H_2S entlang der Rohrwand wird für mehrere verschiedene Einspritzlängen und -geschwindigkeiten berechnet und für diese für niedrig genug befunden, um vor Korrosion zu schützen.

In coal gasification, oxidation and sulfidization cause serious pipe corrosion. This paper attempts to determine the feasibility of reducing such corrosion by injecting steam at pipe entry to modify the boundary-layer gas composition along pipe walls. The injection will form a thin layer on the inner wall, preventing, for a time, contact with the corrosive gases. (Turbulence will eventually force diffusion through the protective layer.) The gas products are assumed to be hydrogen sulfide and steam. The Method of Integral Relations is used to obtain the numerical solutions to the governing equations. With several different injectant lengths and velocities, the concentration of H_2S along the pipe wall is calculated and is found low enough to prevent corrosion.

При газификации угля в результате окисления и образования сульфида возникает сильная коррозия. В данной работе делается попытка определить — существует ли возможность уменьшить эту коррозию с помощью впрыскивания пара во входном участке трубы с целью изменить состав газа в пограничном слое вдоль стенок трубы. Добавление пара способствует образованию тонкого слоя во внутренней стенке, который защищает стенку в течение определённого срока от контакта с вызывающим коррозию газом. (Возможно, что турбулентность ускоряет диффузию через защитный слой). В качестве газа рассматриваются водяной пар и сероводород. Чтобы получить численное решение определяющих уравнений, применяется метод интегральной зависимости. Концентрация сероводорода вдоль стенки трубы вычисляется для нескольких длин и скоростей впрыскивания и считается достаточно малой, чтобы предотвратить коррозию.

Section I: Introduction

In the production of synthetic natural gas (SNG) from coal, the environment in a gasifier contains hydrogen, water, carbon dioxide, carbon monoxide, methane and hydrogen sulfide (see Table 1). The operating temperature is about 1000 °C in all three typical coal gasification processes. The material for all the metallic internal components of the gasifier must be corrosion resistant at temperatures ranging from 700 °C to 1000 °C. We shall take the pipe material to be 310 stainless steel, because this steel is used in existing coal gasifiers. Since hydrogen sulfide is the most corrosive gas and steam is the dominant gas in the pipe mixture, these two gases are considered to be the only fluid components in the mixture.

Table 1. Coal Gasifier Operating Environment

Item	Hygas	Synthane	Consol
Gas Composition:			
In Mole %			
H_2O	24	37	17
CO	7	11	14
CO_2	4	18	6
H_2S	0.8	0.3	0.03
H_2	24	18	45
CH_4	13.3	15	17
N_2	0.3	0.5	0.2
Gas Pressure			
Psig	1100	1000	300
Gas Temperature			
°C	1000	1000	875

There are many ways to inject a protective layer of water along the inner side of pipes, but only two are considered here: (1) water is initially injected at the leading edge in a direction parallel to the axis of the pipe; (2) water is injected at the leading edge for a finite length L in a direction normal to the axis of the pipe. In both cases, the amount of injectant can be small enough to ensure that the contents of the mixture will not be changed greatly.

The effect of strong (i.e., non-viscous) injection on boundary layer characteristics has previously been investigated in [1] and [2]. These theories are adapted to the present problem, to determine the shape of the interface between the pipe flow and the injectant. Viscous diffusion about this interface is then calculated by extending the viscous mixing theory of Chapman [3].

Section 2: Parallel Injection

In this method, water is injected parallel to the axis of a straight circular pipe. To simulate the problem, a two-dimensional free shear model has been used (see Fig. 1). In the upper half-plane, the velocity of the gas mixture is 1524 cm/sec. In the lower halfplane, the velocity of the injectant is taken to be 75 cm/sec and the temperature is taken to be 300 °C, which is the saturated water temperature at 68 atm pressure.

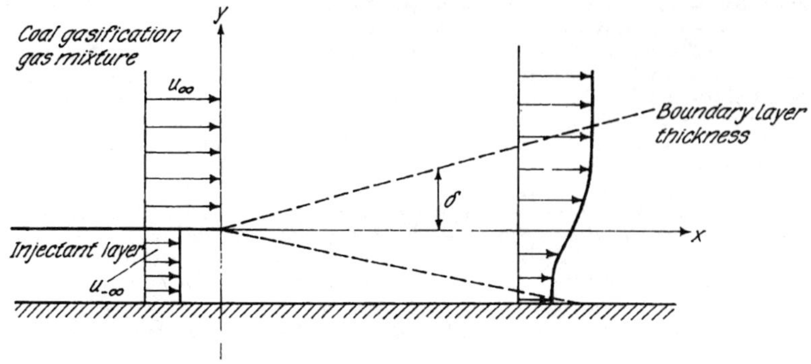

Fig. 1. Coordinate system for parallel injection

The following assumptions are made:

1) The pipe wall in the lower half-plane does not affect the boundary layer growth.
2) The flow is laminar and steady.
3) $Pr = 1$, where $Pr = \nu/\alpha$, α is the heat diffusivity, and ν is the kinematic viscosity of the gas mixture.
4) $Le = 1$, where $Le = \bar{D}/\alpha$ and $\bar{D}$ is the mass diffusion coefficient.
5) $\frac{\bar{\mu}}{\bar{\mu}_\infty} = C(\bar{T}/\bar{T}_\infty)^W$, where $C = 1$ and $W = 1$ in the subsonic case. The viscosity of the gas mixture behaves as $\bar{\mu}/\bar{\mu}_\infty = \bar{T}/\bar{T}_\infty$, where $\bar{\mu}$ is the dynamic viscosity and $\bar{T}$ is the temperature of the gas mixture, and the subscript ∞ denotes the free stream value.
6) The specific heat coefficient, c_p, is constant.
7) The ideal gas law holds.

Governing Equations

Since the Reynolds number is high, the boundary layer equations can be applied:

$$\frac{\partial \bar{\varrho}\bar{u}}{\partial \bar{x}} + \frac{\partial \bar{\varrho}\bar{v}}{\partial \bar{y}} = 0\,, \qquad \bar{\varrho}\bar{u}\frac{\partial \bar{u}}{\partial \bar{x}} + \bar{\varrho}\bar{v}\frac{\partial \bar{u}}{\partial \bar{y}} = \frac{\partial}{\partial \bar{y}}\left(\bar{\mu}\frac{\partial \bar{u}}{\partial \bar{y}}\right), \qquad (2.1),\ (2.2)$$

$$\bar{\varrho}\bar{u}\frac{\partial (c_p\bar{T})}{\partial \bar{x}} + \bar{\varrho}\bar{v}\frac{\partial (c_p\bar{T})}{\partial \bar{y}} = \frac{\partial}{\partial \bar{y}}\left(k\frac{\partial \bar{T}}{\partial \bar{y}}\right); \qquad \bar{\varrho}\bar{u}\frac{\partial \bar{m}}{\partial \bar{x}} + \bar{\varrho}\bar{v}\frac{\partial \bar{m}}{\partial \bar{y}} = \frac{\partial}{\partial \bar{y}}\left(\bar{\varrho}\bar{D}\frac{\partial \bar{m}}{\partial \bar{y}}\right). \qquad (2.3),\ (2.4)$$

In the above, $\bar{u}$ and $\bar{v}$ are the velocity components in the $\bar{x}$ and $\bar{y}$ directions, respectively, $\bar{\varrho}$ is the density, $\bar{m}$ is the mass concentration of each species, and k is the thermal conductivity. With the assumptions that $Pr = Le = 1$ and c_p = constant, the energy equation and species equation can be rewritten as:

$$\bar{\varrho}\bar{u}\frac{\partial \bar{T}}{\partial \bar{x}} + \bar{\varrho}\bar{v}\frac{\partial \bar{T}}{\partial \bar{y}} = \frac{\partial}{\partial \bar{y}}\left(\bar{\mu}\frac{\partial \bar{T}}{\partial \bar{y}}\right), \qquad \bar{\varrho}\bar{u}\frac{\partial \bar{m}}{\partial \bar{x}} + \bar{\varrho}\bar{v}\frac{\partial \bar{m}}{\partial \bar{y}} = \frac{\partial}{\partial \bar{y}}\left(\bar{\mu}\frac{\partial \bar{m}}{\partial \bar{y}}\right). \qquad (2.5\,a),\ (2.5\,b)$$

Introduce the non-dimensional quantities:

$$u = \frac{\bar{u}}{\bar{U}_\infty}, \quad T = \frac{\bar{T}}{\bar{T}_\infty}, \quad x = \frac{\bar{x}}{L}, \quad \varrho = \bar{\varrho}/\bar{\varrho}_\infty\,, \quad \mu = \bar{\mu}/\bar{\mu}_\infty\,, \quad m = \bar{m}\,, \qquad (2.6)$$

$$y = \frac{\bar{y}}{L}\left(\frac{\bar{U}_\infty L}{\nu_\infty}\right)^{1/2}, \qquad v = \frac{\bar{v}}{\bar{U}}\left(\frac{\bar{U}_\infty L}{\nu_\infty}\right)^{1/2},$$

where L is a reference length. The governing equations can be written as:

$$\frac{\partial}{\partial x}(\varrho u) + \frac{\partial}{\partial y}(\varrho v) = 0\,, \qquad \varrho u\frac{\partial u}{\partial x} + \varrho v\frac{\partial u}{\partial y} = \frac{\partial}{\partial y}\left(\mu\frac{\partial u}{\partial y}\right), \qquad (2.7),\ (2.8)$$

$$\varrho u\frac{\partial T}{\partial x} + \varrho v\frac{\partial T}{\partial y} = \frac{\partial}{\partial y}\left(\mu\frac{\partial T}{\partial y}\right), \qquad \varrho u\frac{\partial m}{\partial x} + \varrho v\frac{\partial m}{\partial y} = \frac{\partial}{\partial y}\left(\mu\frac{\partial m}{\partial y}\right). \qquad (2.9),\ (2.10)$$

The boundary conditions are:

$$\text{at}\quad y=\infty\,,\qquad u=1\,,\qquad T=1\,,\qquad m=0.02\,;\qquad\qquad (2.11)$$
$$\text{at}\quad y=-\infty\,,\qquad u=0.05\,,\qquad T=0.45\,,\qquad m=0\,.$$

Under the assumption $Pr = 1$, when applied to the boundary layer equations (as pointed out in [4]), the energy equation can be decoupled from the other equations. Using CROCCO's relation, T is a function of u only, and $T(u) = Au + B$, where A and B are constants determined from the boundary conditions. Similarly, the species equation can also be decoupled by the assumption $Le = 1$ and can be written as $m(u) = A'u + B'$, where A' and B' are constants determined by the boundary conditions.

Applying the boundary conditions (2.11), we get

$$T(u) = 0.5684u + 0.4316\,,\qquad m(u) = 0.0211u - 0.00105\,.\qquad\qquad (2.12a), (2.12b)$$

Once the velocity profile has been determined, the temperature profile and mass concentration profile can also be solved.

To obtain the velocity profile, the continuity equation and momentum equation were solved by the Method of Integral Relations [5]. At first the equations were transformed into incompressible form, by the STEWARTSON transformation:

$$\xi = cx\,,\qquad c = 1 \quad \text{for the subsonic case;}$$

$$\eta = \int_0^y \varrho\,\mathrm{d}y\,,\qquad U = u\,,\qquad \mu = \mu\,,\qquad \varrho = \varrho\,,\qquad \varrho V = Vc - U(\partial\eta/\partial x)_y\,.\qquad\qquad (2.13)$$

Since $\mu = T$, the ideal gas law gives $\varrho T = 1$. The continuity equation and momentum equation become:

$$\frac{\partial U}{\partial \xi} + \frac{\partial V}{\partial \eta} = 0\,,\qquad U\frac{\partial U}{\partial \xi} + V\frac{\partial U}{\partial \eta} = \frac{\partial^2 U}{\partial \eta^2}.\qquad\qquad (2.14), (2.15)$$

Let $f(U)$ be a weighting function. Multiply equation (2.14) by $f(U)$ and equation (2.15) by $f'(U)$, and add the result. We obtain:

$$\frac{\partial}{\partial \xi}(Uf) + \frac{\partial}{\partial \eta}(Vf) = f'\frac{\partial^2 U}{\partial \eta^2}.$$

The equation is now solved separately in the two regions: one is in the upper half-plane, and the other in the lower half-plane.

For the upper half plane, we integrate with respect to η from 0 to ∞. We obtain:

$$\frac{\partial}{\partial \xi}\int_0^\infty fU\,\mathrm{d}\eta + fV\Big|_{\eta=0}^{\eta=\infty} = \int_0^\infty f'(U)\frac{\partial^2 U}{\partial \eta^2}\,\mathrm{d}\eta\,.\qquad\qquad (2.16)$$

Since $V = 0$ at $\eta = 0$, $U = 1$ at $\eta = \infty$, and $f(1) = 0$, the second term on the left-hand side of the above equation becomes zero. We change the variable of integration from η to U and integrate the term on the right-hand side of the equation by parts. We obtain:

$$\frac{\partial}{\partial \xi}\int_{U_0}^1 \Theta U f(U)\,\mathrm{d}U = \frac{f'(U)}{\Theta}\Big|_{U_0}^1 - \int_{U_0}^1 \frac{f''(U)}{\Theta}\,\mathrm{d}U\,,\qquad\qquad (2.17)$$

where

$$\Theta = \frac{1}{\partial U/\partial \eta}\,,\qquad U_0 = U(\eta = 0)\,.$$

Since at $\eta = \infty$, $U = 1$, $\dfrac{\partial U}{\partial \eta} = 0$, $\Theta = \infty$, therefore $\Theta \sim \dfrac{1}{1-U}$.

To avoid a singularity in the equation, we may choose

$$f(U) \sim (1-U)\,,\qquad (1-U)^2\,,\quad \ldots\,, (1-U)^n, \ldots$$

For the first approximation, we take $f(U) = 1 - U$ and $\Theta = \dfrac{1-U_0}{1-U}\,\Theta_0(\xi)$ where

$$\Theta_0(\xi) = \Theta(\xi,\eta)|_{\eta=0}\,.$$

Substitution in equation (2.17) gives

$$\frac{\partial}{\partial \xi}\int_{U_0}^1 (1-U_0)\,\Theta_0 U\,\mathrm{d}U = \frac{f'(U)}{\Theta}\Big|_{U_0}^1 = \frac{1}{\Theta_0}.$$

Since the problem is similar, U_0 = constant, and

$$(1 - U_0)(1 - U_0^2)\frac{d\Theta_0}{d\xi} = \frac{2}{\Theta_0}. \tag{2.18}$$

For the lower half-plane, we use coordinates ξ, $\hat{\eta}$, where $\hat{\eta} = -\eta$, and put a circumflex (^) above each variable. For the first approximation, we take:

$$\hat{f}(\hat{U}) = \hat{U} - 0.05\,, \qquad \hat{\Theta} = \frac{\hat{U}_0 - 0.05}{\hat{U} - 0.05}\hat{\Theta}_0(\xi)\,.$$

Similar to the procedures for the upper half-plane, we obtain:

$$\frac{\partial}{\partial\xi}\int_{\hat{U}_0}^{0.05}(U_0 - 0.05)\,\hat{\Theta}_0\hat{U}\,d\hat{U} = \frac{\hat{f}'(\hat{U})}{\hat{\Theta}_0}\bigg|_{\hat{U}_0}^{0.05} = \frac{-1}{\Theta_0}, \qquad (\hat{U}_0 - 0.05)(0.05^2 - \hat{U}_0^2)\frac{d\hat{\Theta}_0}{d\xi} = \frac{-2}{\hat{\Theta}_0}. \tag{2.19}$$

Applying the matching conditions at $\eta = \hat{\eta} = 0$, we obtain:

$$\Theta_0 = -\hat{\Theta}_0\,, \qquad U_0 = \hat{U}_0\,, \qquad \frac{d\Theta_0}{d\xi} = -\frac{d\hat{\Theta}_0}{d\xi}, \qquad (1 - U_0)(1 - U_0^2) = -(U_0 - 0.05)(0.05^2 - U_0^2) \tag{2.20}$$

Solving numerically for U_0, we find $U_0 = 0.6274$. From similarity we have:

$$\Theta_0 = \beta_0\xi^{1/2}\,, \qquad \hat{\Theta}_0 = \hat{\beta}_0\xi^{1/2}\,,$$

where β_0 and $\hat{\beta}_0$ are constants. Equation (2.18) then becomes:

$$(1 - U_0)(1 - U_0^2)\frac{1}{2}\beta_0\xi^{-1/2} = \frac{2}{\beta_0\xi^{1/2}},$$

or

$$\beta_0 = 2/[(1 - \bar{U}_0)(1 - \bar{U}_0^2)]^{1/2}\,.$$

Recalling

$$\eta = \int_0^y \varrho\,dy\,, \qquad \varrho T = 1 \quad \text{and} \quad U = u\,,$$

we have:

$$y = \int_0^y \frac{d\eta}{\varrho} = \int_{U_0}^U \Theta T\,dU\,, \qquad y = \beta_0(1 - U_0)\,\xi^{1/2}\int_{U_0}^U \frac{AU + B}{1 - U}dU\,,$$

$$\frac{y}{\xi^{1/2}} = \bar{y}\left(\frac{\bar{U}_\infty}{\nu_\infty \bar{X}}\right)^{1/2} = 1.5678\,[0.5684U - \ln(1 - U) - 1.344]\,.$$

Similarly, equation (2.20) becomes:

$$\hat{y} = \xi^{1/2}\hat{\beta}_0(U_0 - 0.05)\int_{U_0}^U \frac{AU + B}{U - 0.05}dU \tag{2.21}$$

or

$$\bar{y}\left(\frac{\bar{U}_\infty}{\nu_\infty \bar{X}}\right)^{1/2} = 2.4299\,[0.5684U + 0.45\ln(U - 0.05) - 0.1095]\,. \tag{2.22}$$

With the velocity profile known, the H_2S concentration profile can be found from equations (2.11) and (2.12). Same calculations were performed with injectant velocity equal to 500 cm/sec. Results are plotted in Figure 2 and Figure 3.

Define the hydrodynamic boundary layer thickness of the lower halfplane as that where the mass concentration of H_2S is 0.0002, which is 99% smaller than that in the coal gasification mixture. Outside the boundary layer, in the lower half-plane, the H_2S concentration is even smaller, and is considered to be low enough to prevent corrosion of the pipes. Then we can say that the pipes are free from corrosion from the entry to the point where the boundary layer thickness meets the wall.

For a large pipe diameter, for example, 90 cm, we can inject large amounts of steam. For a width of injection equal to 1 cm and an injection velocity equal to 75 cm/sec, we find that the boundary layer hits the wall at $X = 19.4$ meters. Therefore, the pipe is protected from corrosion for the initial length of 19.4 m. For an injection velocity equal to 500 cm/sec and a width of injection equal to 1 cm, we find that the initial 31.9 m of the pipe is protected from corrosion. For 90-cm-diameter pipes, *Re* is high and the flow is turbulent. The turbulent model will reduce the protected length considerably.

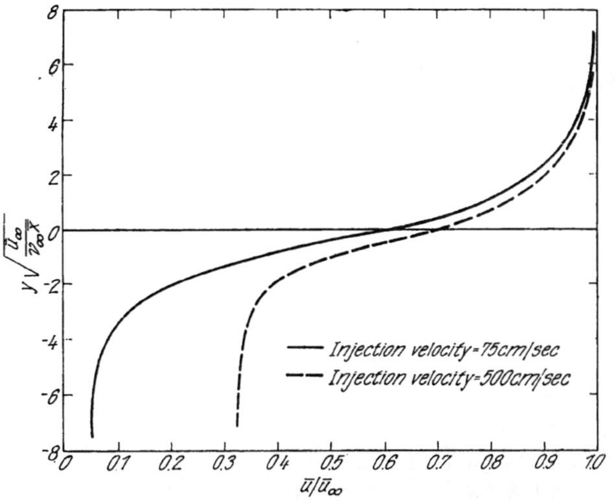

Fig. 2. Numerical solutions of velocity profile for parallel-injection model

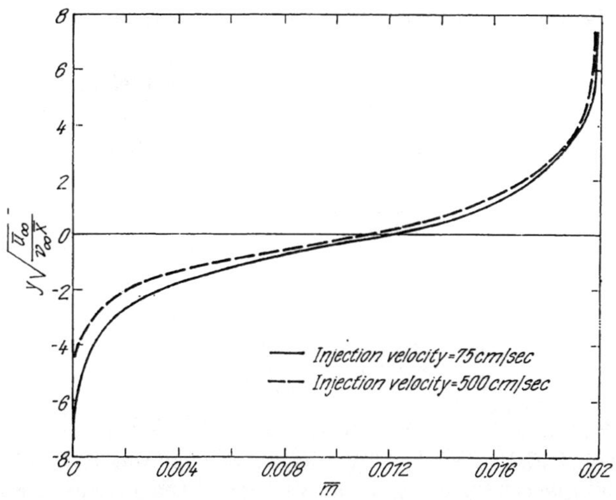

Fig. 3. Numerical solutions of H_2S concentration profile for parallel-injection model

The usual diameter of pipes having serious corrosion in the coal gasification process is 5 cm. Since we want to inject less than 1% of volume of steam into the pipes, we choose an injection velocity of 500 cm/sec and width of injection of 0.3 cm, resulting in the initial 28.7 cm of the pipe being protected from corrosion.

Section III: Normal Injection

In this section, we discuss injecting water normal to the axis of a straight circular pipe. A uniform strong injection is applied over a finite entry length of the pipe. To simplify the problem, the circumferential effect is neglected. A two-dimensional model is used to simulate the problem (see Figure 4). It has been pointed out in [2] that, for strong injection, $V_0/u_\infty \gg Re^{1/2}$, and the injectant layer is inviscid. The viscous effect is confined to a thin layer along the dividing streamline of the injectant and the coal gasification gas mixture. In the first part of this section, we find the dividing streamline, using an inviscid model. In the second part, we find the mass diffusion of H_2S across the dividing streamline by means of the Method of Integral Relations.

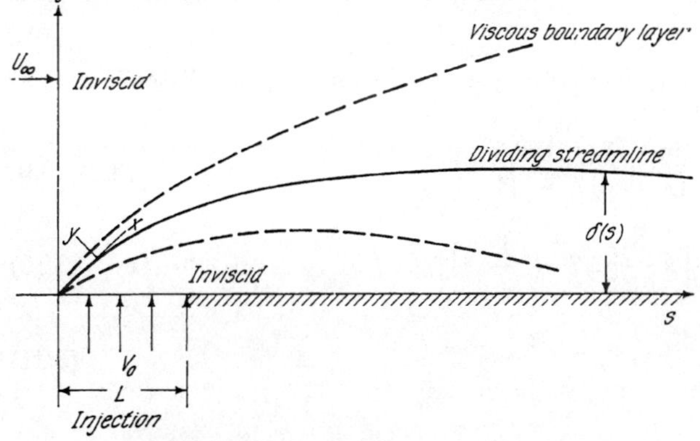

Fig. 4. Coordinate system for normal injection

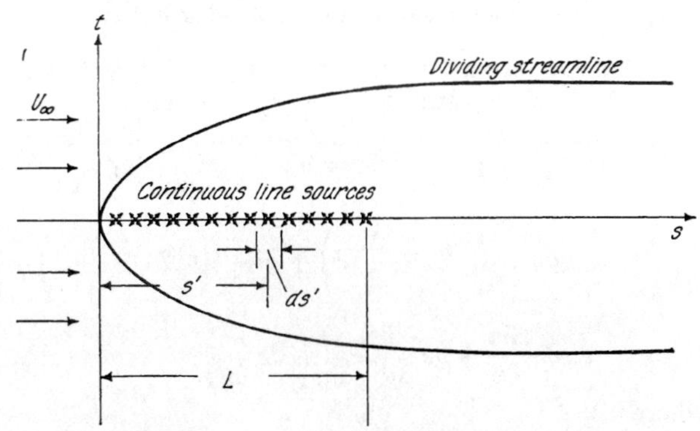

Fig. 5. Diagram for inviscid analysis; uniform flow over a line source

Inviscid Analysis

To find the dividing streamline, a model based on uniform flow over a line source is used to analyze the problem (see Figure 5). A line source of strength σ is located between $s = 0$ and $s = L$, where L is the injection length and s is measured along the streamline. The amount of water injected must be equal to the amount of water coming out from one side of the line source. So $\sigma = 2V_0$, where V_0 is injection velocity. By means of the complex variable $z = s + it$, where t is the transverse coordinate, the complex velocity is given by:

$$u(z) - iv(z) = U_\infty + \int_0^L \frac{\sigma}{2\pi} \frac{ds'}{(z - s')}, \tag{3.1}$$

$$u(z) = \frac{\sigma}{2\pi} \int_0^L \frac{s - s'}{(s - s')^2 + t^2} ds' + U_\infty, \qquad v(z) = \frac{\sigma}{2\pi} \int_0^L \frac{t}{(s - s')^2 + t^2} ds'. \tag{3.2), (3.3}$$

The velocity components of the gas mixture on the dividing streamline are:

$$u(s, \delta) = \frac{V_0}{2\pi} \ln \left[\frac{s^2 + \delta^2}{(s - L)^2 + \delta^2}\right] + U_\infty, \qquad v(s, \delta) = \frac{V_0}{\pi}\left(\tan^{-1}\frac{s}{\delta} - \tan^{-1}\frac{s - L}{\delta}\right), \tag{3.4), (3.5}$$

where δ is the height of the dividing streamline.

Since $\dfrac{d\delta}{ds} = \dfrac{V(s, \delta)}{U(s, \delta)}$, we have:

$$\frac{d\delta}{ds} = \frac{\dfrac{V_0}{\pi}\left(\tan^{-1}\dfrac{s}{\delta} - \tan^{-1}\dfrac{s - L}{\delta}\right)}{\dfrac{V_0}{\pi} \ln \left[\dfrac{s^2 + \delta^2}{(s - L)^2 + \delta^2}\right] + u_\infty}. \tag{3.6}$$

We know that far downstream, the velocity is uniform and equal to U_∞. Hence by conservation of mass, we have $\delta = V_0 L / U_\infty$. With this as the initial condition, the dividing streamline, $\delta(s)$, can be found by integrating equation (3.6) backward along s. The velocity of the gas mixture U_1 along the dividing streamline is given by

$$U_1 = [u(s, \delta)^2 + v(s, \delta)^2]^{1/2}.$$

Across the dividing streamline, the pressure is continuous, but the velocity and the H_2S concentration are discontinuous. Applying BERNOULLI's equation on the dividing streamline, we find the velocity of the injectant U_2 is:

$$U_2 = [V_0^2 + (\varrho_\infty / \hat{\varrho}_\infty)\, U_1^2]^{1/2}. \tag{3.7}$$

Mixing Layer Analysis

To solve for the mass diffusion across the dividing streamline, the following assumptions have been made:

1) Laminar flow.
2) Steady state.
3) $Le = 1$.
4) $Pr = 1$.
5) Since the concentration of H_2S is small in the mixture, we can assume the gas constant R and the specific heats c_p and c_v are all constant in the mixing layer.

Use Cartesian coordinates x and y; x is the distance along the dividing streamline and y is the distance normal to it. The basic governing boundary layer equations are as follows:

$$\frac{\partial}{\partial x}(\bar{\varrho}\bar{u}) + \frac{\partial}{\partial y}(\bar{\varrho}\bar{v}) = 0, \qquad \bar{\varrho}\bar{u}\frac{\partial \bar{u}}{\partial x} + \bar{\varrho}\bar{v}\frac{\partial \bar{u}}{\partial y} = -\frac{dp_e}{dx} + \frac{\partial}{\partial y}\left(\bar{\mu}\frac{\partial \bar{u}}{\partial y}\right), \tag{3.8), (3.9}$$

$$\bar{\varrho}\bar{u}\frac{\partial}{\partial x}\left(c_p \bar{T} + \frac{1}{2}\bar{u}^2\right) + \bar{\varrho}\bar{v}\frac{\partial}{\partial y}\left(c_p \bar{T} + \frac{1}{2}\bar{u}^2\right) = \frac{\partial}{\partial y}\left[\bar{\mu}\frac{\partial}{\partial y}\left(c_p \bar{T} + \frac{1}{2}\bar{u}^2\right)\right], \tag{3.10}$$

$$\bar{\varrho}\bar{u}\frac{\partial \bar{m}}{\partial x} + \bar{\varrho}\bar{v}\frac{\partial \bar{m}}{\partial y} = \frac{\partial}{\partial y}\left(\bar{\varrho}D\frac{\partial \bar{m}}{\partial y}\right), \tag{3.11}$$

where

$$p_e = \bar{\varrho}_\infty R \bar{T}.$$

The boundary conditions are:

$$\bar{u} \to U_1 \quad \text{as} \quad y \to \infty; \qquad \bar{u} \to U_2 \quad \text{as} \quad y \to -\infty.$$

We solve the above equations by the Method of Integral Relations. We solve separately the region above the dividing streamline and the region below it, using the matching conditions along it as boundary conditions.

For the upper region, apply the DORODNITSYN transformation:

$$\xi = \frac{1}{V_m L}\int_0^x \frac{p_e(x)}{p_s} U_1(x)\, dx, \qquad \eta = \frac{U_1(x)}{(\nu_S V_m L)^{1/2}}\int_0^y \frac{\bar{\varrho}}{\varrho_s}\, dy,$$

$$v = \frac{V_m L u \dfrac{\partial \eta}{\partial x}}{U_1 p_e / p_s} + \frac{(V_m L)^{1/2}}{U_1 (\nu_s)^{1/2}} \bar{v} \frac{T_s}{\bar{T}}, \qquad u = \frac{\bar{u}}{U_1}, \qquad 1 - h = \frac{c_p \bar{T} + \dfrac{1}{2}\bar{u}^2}{c_p T_s}, \qquad m = \bar{m} - \bar{m}_\infty. \tag{3.12}$$

Here subscript "s" represents quantities at the stagnation point and $\frac{1}{2} V_m^2 = c_p \overline{T}_s$, while L is some reference length.

The resulting governing equations become:

$$\frac{\partial u}{\partial \xi} + \frac{\partial v}{\partial \eta} = 0\,, \qquad u\frac{\partial u}{\partial \xi} + v\frac{\partial u}{\partial \eta} = \beta(1 - h - u^2) + \frac{\partial^2 u}{\partial \eta^2}, \tag{3.13), (3.14}$$

$$u\frac{\partial h}{\partial \xi} + v\frac{\partial h}{\partial \eta} = \frac{\partial^2 h}{\partial \eta^2}, \qquad u\frac{\partial m}{\partial \xi} + v\frac{\partial m}{\partial \eta} = \frac{\partial^2 m}{\partial \eta^2}, \tag{3.15), (3.16}$$

where

$$\beta = \frac{\alpha_e}{\alpha_e(1 - \alpha_e)}, \qquad \alpha_e = \frac{U_1(x)}{V_m}.$$

Let $f(u)$ and $g(u)$ be weighting functions. Multiply equation (3.13) by $f(u)$ and equation (3.14) by $f'(u)$, and add the result. We obtain:

$$\frac{\partial}{\partial \xi}(uf) + \frac{\partial}{\partial \eta}(vf) = \beta f'(1 - h - u^2) + f'\frac{\partial^2 u}{\partial \eta^2}. \tag{3.17}$$

Multiply equation (3.13) by $hg(u)$, equation (3.14) by $hg'(u)$, equation (3.15) by $g(u)$, and add the result. We obtain:

$$\frac{\partial}{\partial \xi}(hug) + \frac{\partial}{\partial \eta}(hvg) = \beta(1 - h - u^2)\, hg'(u) + \frac{\partial^2 u}{\partial \eta^2} hg'(u) + g\frac{\partial^2 h}{\partial \eta^2}. \tag{3.18}$$

Integrate the above three equations with respect to η from 0 to ∞ and then change the variable of integration from η to u. We get:

$$\frac{\partial}{\partial \xi}\int_{u_0}^{1} \Theta u f(u)\, du = \beta \int_{u_0}^{1} f'(u)\,(1 - u^2)\,\Theta\, du - \beta\int_{u_0}^{1} f'(u)\,\Theta h\, du + \frac{f'(u)}{\Theta}\bigg|_{u_0}^{1} - \int_{u_0}^{1}\frac{f''(u)}{\Theta}\, du - fv\bigg|_{\eta=0}^{\eta=\infty}, \tag{3.20}$$

$$\begin{aligned}\frac{\partial}{\partial \xi}\int_{u_0}^{1} hug\Theta\, du = {}&\beta\int_{u_0}^{1}(1 - u^2)\, hg'\Theta\, du - \beta\int_{u_0}^{1} h^2 g'\Theta\, du + \frac{1}{\Theta} hg'\bigg|_{u_0}^{1} - \\ &- \int_{u_0}^{1}\frac{1}{\Theta}\left(g''h + g'\frac{\partial h}{\partial u}\right) du + \frac{g}{\Theta}\frac{\partial h}{\partial u}\bigg|_{u_0}^{1} - \int_{u_0}^{1} g'\frac{\partial h}{\partial u}\frac{1}{\Theta}\, du - hvg\bigg|_{\eta=0}^{\eta=\infty},\end{aligned} \tag{3.21}$$

$$\begin{aligned}\frac{\partial}{\partial \xi}\int_{u_0}^{1} mug\Theta\, du = {}&\beta\int_{u_0}^{1}(1 - u^2)\, mg'\Theta\, du - \beta\int_{u_0}^{1} mhg'\Theta\, du + \frac{1}{\Theta} mg'\bigg|_{u_0}^{1} - \\ &- \int_{u_0}^{1}\frac{1}{\Theta}\left((g''m + g'\frac{\partial m}{\partial u}\right) du + \frac{g}{\Theta}\frac{\partial m}{\partial u}\bigg|_{u_0}^{1} - \int_{u_0}^{1} g'\frac{\partial m}{\partial u}\frac{1}{\Theta}\, du - mvg\bigg|_{\eta=0}^{\eta=\infty},\end{aligned} \tag{3.22}$$

where

$$\Theta = \frac{1}{\partial u/\partial \eta}, \qquad u_0 = u(\xi, \eta)_{\eta=0}\,.$$

For the first approximation, we choose:

$$f(u) = 1 - u\,, \qquad g(u) = 1\,,$$

$$\Theta = \frac{1 - u_0}{1 - u}\Theta_0(\xi)\,, \qquad K = K_0(\xi) = h\Theta\,, \qquad J = J_0(\xi) = m\Theta\,.$$

Then equations (3.20), (3.21) and (3.22) respectively become:

$$\frac{\partial}{\partial \xi}\left[\frac{\Theta_0}{2}(1 - u_0)\,(1 - u_0^2)\right] = \beta(1 - u_0)\,\Theta_0\left(u_0 + \frac{u_0^2}{2} - \frac{3}{2}\right) + \beta K_0(1 - u_0) + \frac{1}{\Theta_0}, \tag{3.23}$$

$$\frac{\partial}{\partial \xi}[J_0(1 - u_0^2)] = \frac{2J_0}{\Theta_0^2(1 - u_0)}, \qquad \frac{\partial}{\partial \xi}[K_0(1 - u_0^2)] = \frac{2K_0}{\Theta_0^2(1 - u_0)}, \tag{3.24), (3.25}$$

For the lower region, we put circumflex (^) above all variables as before, and take the DORODNITSYN transformation:

$$\hat{\xi} = \frac{1}{\hat{V}_m L} \int_0^x \frac{p_e(x)}{\hat{p}_s} U_2(x)\, dx; \qquad \hat{p}_e(x) = p_e(x)\,,$$

$$\hat{\eta} = \frac{U_2(x)}{(\nu_S \hat{V}_m L)^{1/2}} \int_0^y \frac{\bar{\varrho}}{\hat{\varrho}_s} dy\,, \qquad \hat{v} = \frac{\hat{V}_m L U_2 \frac{\partial \hat{\eta}}{\partial x}}{U_2 p_e/p_s} + \frac{(V_m L)^{1/2}}{U_2(\nu_s)^{1/2}} \hat{\bar{v}} \frac{\hat{T}_s}{\hat{T}}\,,$$

$$\hat{u} = \frac{\bar{u}}{U_2}, \qquad \hat{m} = \hat{\bar{m}}\,,$$

Using the same procedures that lead to equation (3.23), we get the following integral relation for the lower region:

$$\frac{\partial}{\partial \hat{\xi}}\left[\frac{\hat{\Theta}_0}{2}(1-\hat{u}_0)(1-\hat{u}_0^2)\right] = \hat{\beta}(1-\hat{u}_0)\,\hat{\Theta}_0\left(\hat{u}_0 + \frac{u_0^2}{2} - \frac{3}{2}\right) + \hat{\beta}\hat{K}_0(1-\hat{u}_0) + \frac{1}{\hat{\Theta}_0}. \tag{3.26}$$

Since u, Θ, T and m are continuous across the dividing streamline, we have four matching conditions. Together with the four integral relations (3.23), (3.24), (3.25), and (3.26), we have eight equations to solve for eight unknowns.

The four matching conditions are:

$$\hat{u}_0 = u_0 U_1/U_2\,, \qquad \hat{\Theta}_0 = -\Theta_0\left(\frac{V_m}{\hat{V}_m}\right)^{1/2}\left(\frac{U_2}{U_1}\right)^2, \tag{3.27), (3.28}$$

$$K_0 = -\Theta_0\left(\frac{V_m}{\hat{V}_m}\right)^{1/2}\left(\frac{U_2}{U_1}\right)^2 \frac{\hat{T}_s - T_s + K_0 T_s/\Theta_0}{\hat{T}_s}\,, \tag{3.29}$$

$$m = \left(\bar{m}_\infty - \frac{J_0}{\Theta_0}\right)\left(\frac{1-\hat{u}}{1-\hat{u}_0}\right). \tag{3.30}$$

Substituting these matching conditions into the equations (3.23), (3.24), (3.25), and (3.26) to eliminate the unknowns $\hat{\Theta}_0$, $\hat{U}_0$, $\hat{K}_0$, we obtain the following:

$$\begin{bmatrix} A_{11} & A_{12} & 0 & 0 \\ A_{21} & A_{22} & 0 & 0 \\ 0 & A_{32} & A_{33} & 0 \\ 0 & A_{42} & 0 & A_{44} \end{bmatrix} \begin{bmatrix} \dot{\Theta}_0 \\ \dot{U}_0 \\ \dot{J}_0 \\ \dot{K}_0 \end{bmatrix} = \begin{bmatrix} B_1 \\ B_2 \\ B_3 \\ B_4 \end{bmatrix}, \tag{3.31}$$

where

$$A_{11} = (1-u_0)(1-u_0^2)\,, \qquad A_{12} = (3u_0+1)(u_0-1)\,,$$

$$A_{21} = (1-\alpha u_0)(1-\alpha^2 u_0^2)/\alpha^2\,, \qquad A_{22} = \left(-\frac{1}{\alpha} - 2u_0 + 3\alpha u_0^2\right)\Big/\Theta_0\,,$$

$$A_{32} = -2u_0 J_0\,, \qquad A_{33} = 1 - u_0^2\,,$$

$$A_{42} = -2u_0 K_0\,, \qquad A_{44} = 1 - u_0^2\,,$$

$$B_1 = \beta(1-u_0)(u_0+3)(u_0-1)\,\Theta_0 + 2\beta K_0(1-u_0) + \frac{2}{\Theta_0}, \qquad B_2 = \frac{2J_0}{\Theta_0^2(1-u_0)},$$

$$B_3 = \frac{2K_0}{\Theta_0^2(1-u_0)},$$

$$B_4 = -\Theta_0\left(-\frac{2}{\alpha^3} + \frac{u_0}{\alpha^2} + u_0^3\right)\frac{d\alpha}{d\xi} + \frac{2\alpha}{\Theta_0} + \\ + \frac{2\Theta_0 V_m}{\alpha^2 \hat{V}_m}\hat{\beta}(1-\alpha u_0)\left(\alpha u_0 - \frac{\alpha^2 u_0^2}{2} - \frac{1}{2} - \frac{T_s}{\hat{T}_s} - \frac{K_0 T_s}{\Theta_0 \hat{T}_s}\right);$$

$$\alpha = U_1/U_2\,, \qquad \hat{\beta} = \frac{1}{\hat{\alpha}_e(1-\hat{\alpha}_e)}\frac{d\hat{\alpha}_e}{d\hat{\xi}} = \frac{\hat{V}_m}{V_m}\frac{U_1}{U_2}\frac{\hat{p}_s}{p_s}\frac{1}{\alpha_e(1-\hat{\alpha}_e)}\frac{d\hat{\alpha}_e}{d\xi}, \qquad \hat{\alpha}_e = u_2/\hat{V}_m\,.$$

Initial conditions are needed to solve the above system of differential equations (3.31). Since at $\xi = 0$, $\Theta_0 = 0$, we have a singularity at $\xi = 0$. In order to avoid the singularity point, we integrate the equations starting at $\xi = \varepsilon$, where ε is a small positive number. From the inviscid model, $U_1(0)$ and $U_2(0)$ are found. Assuming that the pressure gradient has no effect in the vicinity of the leading edge, then the technique discussed in Section 2 can be applied to solve for Θ_0, U_0, J_0 and K_0 at $\xi = \varepsilon$. We take these as the initial conditions and solve the system of differential equations numerically. With Θ_0, U_0, J_0 and K_0 determined for general ξ, the velocity, temperature and H_2S concentration can be found.

Define the boundary layer thickness of the lower half-plane as the normal distance where the H_2S concentration is 0.0002, which is 1% of that in the gas mixture. For comparison with the parallel injection model, we use the same pipe diameter of 5 cm and the same amount of steam, for an injection length of 5 cm and injection velocity

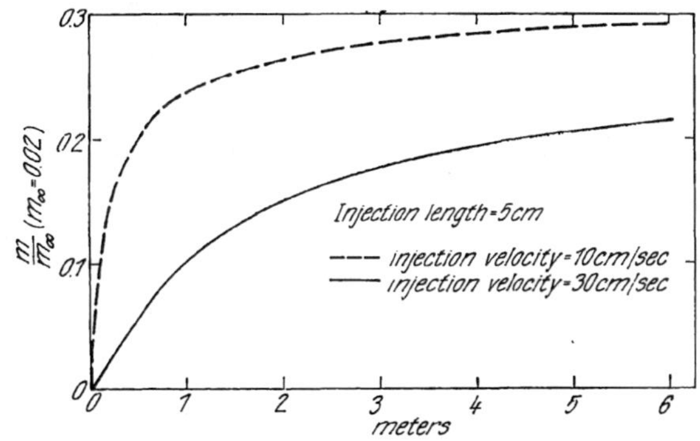

Fig. 6. Numerical solutions of H_2S concentration at the wall of pipe for normal-injection model

of 30 cm/sec. We find that the boundary layer hits the wall at a position 17 cm from the leading edge. The concentration of H_2S at the wall is calculated and plotted in Figure 6.

Section IV. Conclusion

The Method of Integral Relations has been applied successfully to solve the present mixing problem. The accuracy of the solution increases with succeeding orders of approximation, but the calculations are increasingly complicated. In this paper only the first order of approximation is used.

For large pipe diameter, say 1 meter, we can inject a layer 1 cm thick. From the parallel injection analysis, an initial pipe length of approximately 32 meters is found to be protected from corrosion. For small pipe diameter, say 5 cm, the usual pipe diameter in present coal gasification processes, only a thin layer can be injected. From the calculations in Section III, the concentration of H_2S at the wall of the pipes was found to be reduced approximately 70 percent for an initial length of 6 m.

Acknowledgments

One author (W.C.-C.Y.) wishes to express his gratitude for the encouragement given him by A. V. LEVY, program advisor in the Materials and Molecular Research Division, Lawrence Berkeley Laboratory.

This work was supported by the Assistant Secretary for Materials Sciences, Office of Basic Energy Sciences, U.S. Department of Energy, under Contract W-7405-ENG-48.

List of Symbols

A_{ij} matrix elements
c_p, c_v specific heats
$\bar{D}$ molecular diffusion coefficient
D diffusivity of H_2S
f, g weighting functions
h dimensionless enthalpy
J $j\Theta$
k thermal conductivity
K $m\Theta$
L characteristic length
Le Lewis number
$\bar{m}, m$ species mass concentration
p pressure
Pr Prandtl number
R gas constant
Re Reynolds numbers
t transverse coordinate
$\bar{T}, T$ temperatures
$\bar{u}, \bar{v}$ velocity components (dimensional)
u, v velocity components (dimensionless)
$\bar{U}_\infty$ free stream velocity
U_1 velocity along dividing streamline
U, V transformed velocity components
V_0 injection velocity
$\bar{x}, \bar{y}$ Cartesian coordinates (dimensional)
x, y Cartesian coordinates (dimensionless)
z $s + it$, complex variable in inviscid flow
α, β functions of velocity at outer edge of mixing layer
η $\int_0^y \varrho \, dy$
Θ $1/(\partial u/\partial \eta)$
$\bar{\mu}, \mu$ coefficients of viscosity
$\bar{\nu}, \nu$ coefficients of kinematic viscosity
ν_∞ free stream value
ξ cx, coordinate
$\bar{\varrho}, \varrho$ densities
σ line source strength

Diacritical Symbols

$\bar{\ }$ dimensional quantities
$\hat{\ }$ variables in lower half of mixing region

Subscripts

e value at outer edge of mixing layer
0 quantities at $\eta = 0$, mixing layer interface
s quantities at stagnation point
∞ free stream value

Literature

1 COLE, J. D.; AROESTY, J., The Blowhard Problem — inviscid flows with surface injection, J. Heat Transfer, 2, 1167—1183 (1968).
2 SMITH, F. T.; STEWARTSON, K., Plate Injection into a separated supersonic boundary layer, J. Fluid Mech., 58 (Part I), 143—159 (1973).
3 CHAPMAN, D. R., Laminar mixing of a compressible fluid, National Advisory Committee for Aeronautics Report 958 (1950).
4 SCHLICHTING, H., Boundary layer theory, 6th edition, McGraw-Hill, New York, 1968.
5 HOLT, M., Numerical methods in fluid dynamics, Springer-Verlag, Berlin, Heidelberg, New York, 1977.

Eingereicht am 13. 4. 1981

Anschrift: WILLIAM CHOR-CHUN YEUNG / MAURICE HOLT, Lawrence Berkeley Laboratory, University of California, Berkeley, Ca. 94720

KLEINE MITTEILUNGEN

ZAMM 62, 401 – 402 (1982)

W. Meyer zur Capellen

Die Wendekurven der sphärischen Kinematik und die elliptischen Integrale

Die Projektion der Wendekurve der sphärischen Kinematik [1] auf die zur momentanen Drehachse senkrechte x,y-Ebene hat mit R als Radius der „Einheitskugel" die Polargleichung

$$r = R \sin \lambda(\varphi)\,, \tag{1}$$

wobei der Winkel λ aus

$$\sin \lambda \cos \lambda = \tan \vartheta \sin \varphi \tag{2}$$

folgt mit ϑ als gewissem Parameter. Hiernach ergibt sich auch mit der Abkürzung $\nu = 2 \tan \vartheta$ für die jeweils zwei Radienvektoren r_1 und r_2 bei gegebenem Winkel (vgl. a. Bild 1 a/c)

$$r^2 = \tfrac{1}{2} R^2 \left[1 \pm \sqrt{1 - \nu^2 \sin^2 \varphi}\right]. \tag{3}$$

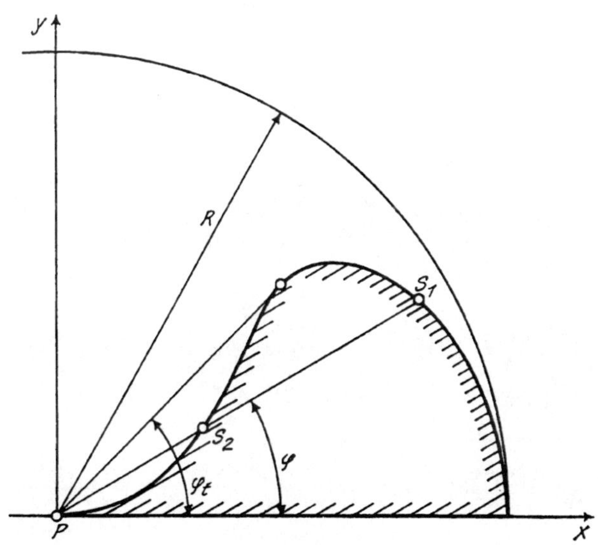

Bild 1a

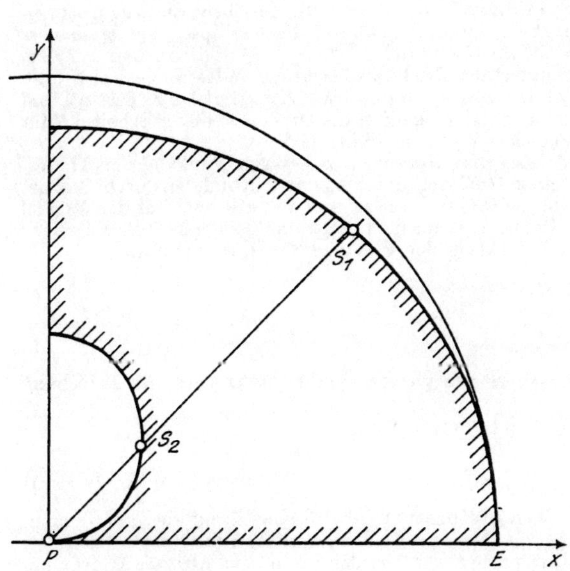

Bild 1b

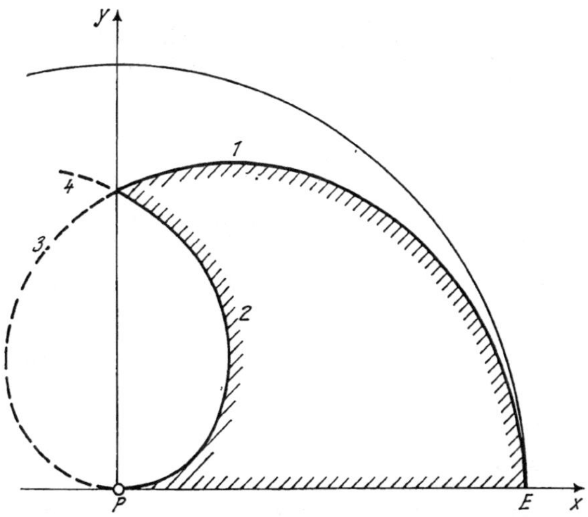

Bild 1c

Es sei nun der Inhalt A der zwischen den beiden Kurventeilen liegenden Fläche gesucht (vgl. Bild 1 a/c). Hierfür folgt dann

$$A = \tfrac{1}{2} \int r^2 \,\mathrm{d}\varphi = \tfrac{1}{2} \int r_1^2 \,\mathrm{d}\varphi - \tfrac{1}{2} \int r_2^2 \,\mathrm{d}\varphi =$$

$$= \tfrac{1}{2} R^2 \int\limits_{\varphi_a}^{\varphi_b} \sqrt{1 - \nu^2 \sin^2 \varphi}\,\mathrm{d}\varphi \tag{4}$$

mit der unteren Grenze $\varphi_a = 0$ und der noch zu bestimmenden oberen Grenze φ_b. Hierbei sind drei Fälle zu unterscheiden:

1. Fall: Es sei $\tan \vartheta > 1/2$ oder $\nu > 1$ (vgl. Bild 1a). Hier fallen die Werte r_1 und r_2 für den Winkel $\varphi = \varphi_t$, der oberen Grenze φ_b, zusammen, wobei, wenn die Zahl $k = 1/\nu$ eingeführt wird, φ_t aus $\sin \varphi_t = 1/\nu = k$ folgt.

Dann führt die Substitution

$$\nu \sin \varphi = \sin \psi \quad \text{oder} \quad \sin \varphi = k \sin \psi \tag{5}$$

mit $\cos \varphi \,\mathrm{d}\varphi = k \cos \psi \,\mathrm{d}\psi$, und wenn

$$\cos \varphi = \sqrt{1 - k^2 \sin^2 \psi} = \Delta \tag{6}$$

gesetzt wird, sowie

$$\cos^2 \psi = [\Delta^2 - (1 - k^2)]/k^2$$

und den Grenzen $\psi_a = 0$, $\psi_b = \pi/2$ schließlich, vgl. z. B. [2], auf

$$A = \frac{k}{2} R^2 \int\limits_0^{\pi/2} \frac{\cos^2 \psi}{\Delta} \,\mathrm{d}\psi = \frac{R^2}{2k} \left[\int\limits_0^{\pi/2} \Delta \,\mathrm{d}\psi - \int\limits_0^{\pi/2} \frac{1 - k^2}{\Delta} \,\mathrm{d}\psi\right] \tag{8a}$$

oder auf

$$A = R^2 \tan \vartheta' \left[E\left(k, \frac{\pi}{2}\right) - \cos^2 \varphi_t F\left(k, \frac{\pi}{2}\right)\right] \tag{8b}$$

mit F als dem vollständigen elliptischen Integral erster und E als dem vollständigen elliptischen Integral zweiter Gattung. Der Parameter k wird üblicherweise gleich $\sin \alpha$ gesetzt, d. h. hier ist $\alpha = \varphi_t$!

2. Fall: Es sei $\tan \vartheta > 1/2$ (Bild 1b). Jetzt ergeben sich zwei getrennte Kurventeile. Die Gleichung (4) bleibt erhalten, nur bleibt jetzt $\nu < 1$, und die obere Grenze von φ ist hier $\varphi = \pi/2$. Damit ergibt sich aber sofort

$$A = \frac{1}{2} R^2 E\left(\nu, \frac{\pi}{2}\right). \tag{9}$$

3. Fall: Es sei jetzt $\tan \vartheta = 1/2$, d. h. $\nu = 1$. Dann fallen die Werte r_1 und r_2 für $\varphi = \varphi_t = \pi/2$ zusammen (vgl. Bild 1c), und es ergibt sich aus den vorstehenden Gleichungen, aber auch unmittelbar

$$A = \tfrac{1}{2} R^2 \int\limits_0^{\pi/2} \cos \varphi \,\mathrm{d}\varphi = \tfrac{1}{2} R^2\,. \tag{10}$$

Hier sei noch auf das Folgende hingewiesen: In diesem Sonderfall wird $r_1 = R \cos \varphi/2$, $r_2 = R \sin \varphi/2$, und, berechnet man den Inhalt der Fläche unter dem Kurvenzug 1, 3 und der x-Achse, so wird

$$A = \frac{1}{2} R^2 \int_0^\pi \left(\cos \frac{\varphi}{2}\right)^2 d\varphi = \frac{\pi}{4} R^2 . \tag{11}$$

Würde man mit einem Planimeter den Kurvenzug 1, 3, 2, 4, x-Achse befahren, so würde man den Flächeninhalt des Halbkreises erhalten.

Zieht man durch P einen Strahl unter dem Winkel φ, so ist, da nach Gleichung 3 doch $r_1^2 + r_2^2 = R^2$ ist (vgl. a. Anm. 1), die vom Großkreis, vom Strahl φ und vom Kurventeil r_1 begrenzte Fläche gleich der vom Strahl φ und dem Kurventeil r_2 begrenzten Fläche.

Literatur

1 MEYER ZUR CAPELLEN, W., Die sphärischen Analoga zu den ϱ-Kurven der ebenen Kinematik von H. ALT, Mechanism and Machine Theory 17 (1982).

2 MEYER ZUR CAPELLEN, W., Integraltafeln. Springer-Verlag. Berlin-Göttingen-Heidelberg 1950.

Eingereicht am 20. 11. 1980

Anschrift: o. Prof. em. Dr.-Ing. WALTHER MEYER ZUR CAPELLEN, Ronheider Weg 46, D-5100 Aachen, BRD

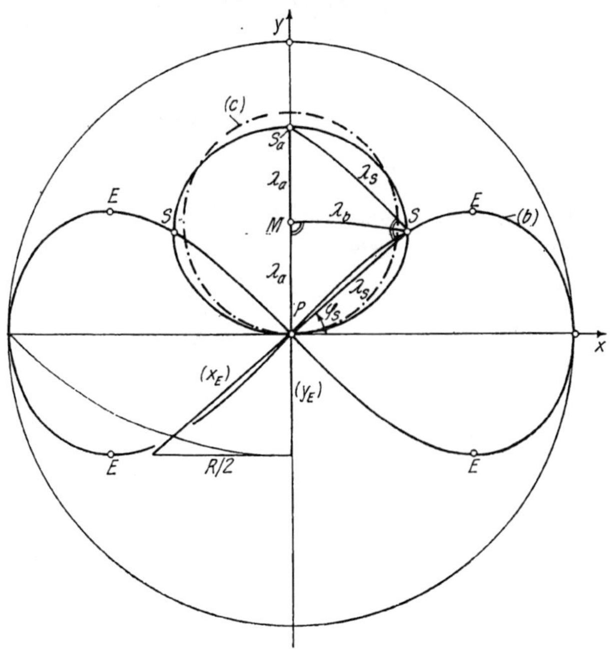

Bild 1a.

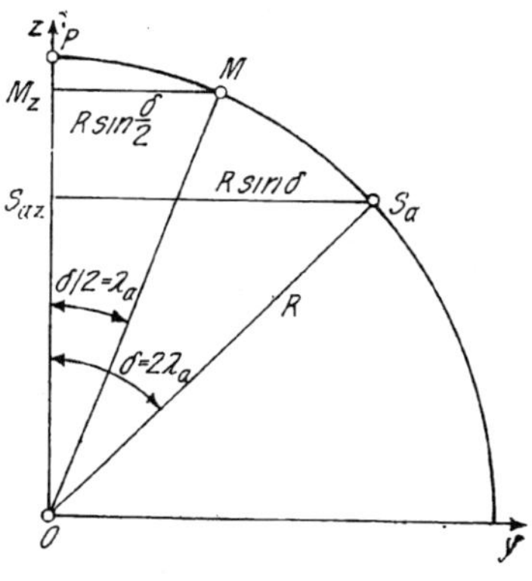

Bild 1b.

ZAMM 62, 402–404 (1982)

W. MEYER ZUR CAPELLEN

Die Scheitelkurve sphärischer Thalesellipsen

a. Die Thalesellipse

In der Kinematik sphärischer Bewegungen d. h. der Bewegungen um einen festen Punkt, begegnet man wiederholt den sphärischen Thalesellipsen als Ortskurven [1, 2]. Sie können als Durchdringungskurven schiefer Kreiskegel mit der Einheitskugel vom Mittelpunkt O (und vom Radius R bei Maßbetrachtungen) aufgefaßt werden. In sphärischen Polarkoordinaten ist ihre Gleichung durch

$$\tan \lambda = \tan \delta \sin \varphi \tag{1}$$

gegeben mit λ als „geographischer Breite", mit φ als „geographischer Länge" und mit dem Winkel δ als Parameter (vgl. Bild 1a, b).

Legt man in P, dem „Nordpol", an die Kugel eine Tangentialebene, so wird diese in einem Kreis vom Durchmesser $d = R \tan \delta$ mit der Polargleichung $r_T = d \sin \varphi$ geschnitten.

In Bild 1a stellt die x,z-Ebene die Tangentialebene an alle Kegel mit der z-Achse als gemeinsamer Mantellinie der Kegel dar, und Bild 1b zeigt die Projektion der sphärischen Thalesellipse (Kurve „a") auf die x,y-Ebene, wobei diese die Polargleichung

$$r = R \sin \lambda(\varphi) \tag{2}$$

hat mit λ aus Gleichung (1).

b. Die Scheitelkurve

Auf dem Großkreis, in welchem die z,y-Ebene die Kugel schneidet, liegt die eine „Achse" der Thalesellipse, ihre Halbachse $PM = MS_a = \lambda_a$ ist gleich $\delta/2$, (vgl. a. Bild 1b), und auf dem dazu senkrechten Großkreis liegen die anderen Scheitel S (vgl. Bild 1a). Die zweite „Halbachse" MS sei gleich λ_b gesetzt, und es interessiert im Folgenden die Ortskurve dieser Scheitel S für alle möglichen Werte des Winkels δ.

Zieht man nun von einem beliebigen Punkt C der Thalesellipse einen Großkreis durch S_a und den anderen durch P, so ist der Winkel PCS_a ein rechter, und dies gilt auch für den Winkel PSS_a. Damit folgt aus dem rechtwinkligen sphärischen Dreieck PSS_a (Bild 1a) für den Winkel $PS = \lambda_s$ sehr einfach

$$\cos \delta = \cos^2 \lambda_s$$

oder

$$\cos \lambda_s = \sqrt{\cos \delta} . \tag{3}$$

Ebenso liefert das gleiche sphärische Dreieck die Beziehung

$$\cot^2 \left(\frac{\pi}{2} - \varphi_s\right) = \cos \delta, \text{ d. h.}$$

$$\tan \varphi_s = \sqrt{\cos \delta} . \tag{4}$$

Für einen gegebenen Winkel δ sind damit die Koordinaten des Scheitels (bzw. der beiden Scheitel) S festgelegt. Auch hat, wie leicht nachgeprüft werden kann, die Abszisse x dort ein Extremum.

Eliminiert man nun aus den Gleichungen (3) und (4) den Winkel δ, so folgt als Polargleichung der Scheitelkurve (Index s fortgelassen)

$$\lambda = \operatorname{arc}\cos(\tan\varphi) \tag{5a}$$

oder

$$\cos\lambda = \tan\varphi\,. \tag{5b}$$

Die vom Kugelmittelpunkt O zu den einzelnen Punkten dieser Ortskurve gezogenen Strahlen bilden einen Kegel vierter Ordnung.

Für die Projektion auf die x,y-Ebene erhält man in Polarform

$$r = R\sin\lambda = R\sqrt{1-\tan^2\varphi} \tag{6}$$

als Kurve vierter Ordnung (vgl. Kurve (b) in Bild 1a). In Parameterform gilt auch

$$x = r\cos\varphi = R\sqrt{\cos 2\varphi}\,,\qquad y = x\tan\varphi\,,$$

$$z = R\cos\lambda = R\tan\varphi\,. \tag{7 a/c}$$

Extrema der Koordinate y ergeben sich, wie eine einfache Rechnung zeigt, für

$$\tan\varphi_E = \sqrt{\sqrt{2}-1}\,,\qquad r_E = R\sqrt{2-\sqrt{2}} \tag{8 a, b}$$

bei einem Winkel δ_E, der aus $\cos\delta_E = \sqrt{2}-1$ folgt. Ferner gilt

$$|x_E| = R\sqrt{\sqrt{2}-1} = R\tan\varphi_E = z_E\,,$$

$$|y_E| = R(\sqrt{2}-1) = R\cos\delta_E\,. \tag{9a/c}$$

Da $x_E^2 = R\cdot|y_E|$ ist, kann außer y_E auch x_E geometrisch leicht gewonnen werden (vgl. Bild 1a, dritter Quadrant).

c. Krümmungen

Im Punkt $\varphi = 0$ und $\lambda = \pi/2$ ist $dz/dy = \pm 1$, d. h. die Schmiegungsebenen der sphärischen Kurve bilden mit der x,y-Ebene einen Winkel von $\pm 45°$ (die Projektion dieser sphärischen Kurve auf die y,z-Ebene liefert den gleichen Verlauf wie die Projektion auf die x,y-Ebene). Nun sind nach einer früher entwickelten Formel [1] die in den Schmiegungsebenen liegenden Krümmungskreise hier identisch mit den Großkreisen. Diese projizieren sich aber auf die x,y-Ebene als Ellipse mit den Halbachsen $a = R$ und $b = R\cos 45°$, und der Krümmungsradius dieser Ellipse an der betrachteten Stelle ist auch der Krümmungsradius der Projektion der sphärischen Kurven, d. h. es ist $\varrho = b^2/a = R/2$.

Beim Übergang zur ebenen Bewegung geht beiläufig die Scheitelkurve in die Geraden $y = \pm x$ über, d. h. trivial in die Verbindungsgeraden der Endpunkte der zur x-Achse parallelen Durchmesser der oben genannten Kreise vom Durchmesser d.

d. Flächeninhalte

Interessant sind noch einige Bemerkungen über gewisse Flächeninhalte.

d.1. Die Scheitelkurve

d.1.a. *Oberfläche*: Es sei die von der Scheitelkurve auf der Kugel umschlossene Fläche A in den Grenzen $\varphi_1 = -\pi/4$ und $\varphi_2 = \pi/4$ gesucht. Das Flächenelement ist durch $dA = R\sin\lambda\, d\lambda\, R\, d\varphi$ bestimmt, d. h. es wird

$$A = 2R^2\int_0^{\pi/4}(1-\cos\lambda)\,d\varphi = 2R^2\int_0^{\pi/4}(1-\tan\varphi)\,d\varphi =$$

$$= R^2(\pi/2 + 2\ln\cos\varphi)$$

oder

$$A = R^2(\pi/2 - \ln 2)\,. \tag{10}$$

d.1.b. *Projektion auf die x,y-Ebene*: Hier wird

$$A = \tfrac{1}{2}\int r^2\,d\varphi = \tfrac{1}{2}R^2\int\sin^2\lambda\,d\varphi = R^2\int_0^{\pi/4}(1-\tan^2\varphi)\,d\varphi =$$

$$= R^2\int_0^{\pi/4}[2-(1+\tan^2\varphi)]\,d\varphi = R^2(2\varphi-\tan\varphi)_0^{\pi/4}$$

oder

$$A = R^2\pi/2 - R^2\,. \tag{11}$$

Dieser Flächeninhalt ist also gleich der schraffierten Fläche in Bild 2. Danach ist aber auch der Inhalt der Fläche, welche von der y-Achse, dem Großkreis $z = 0$ und der Scheitelkurve begrenzt wird, gerade gleich R^2!

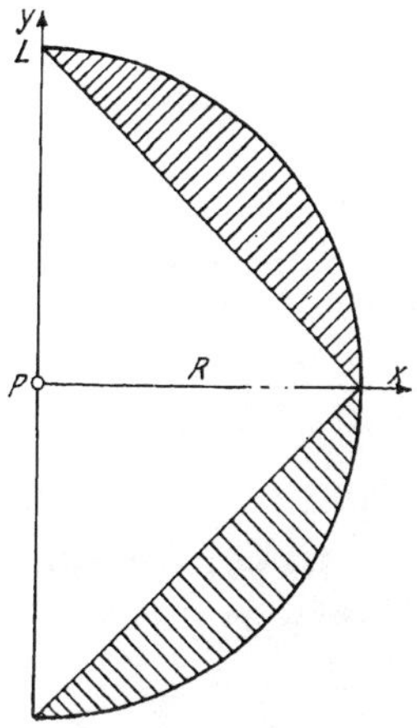

Bild 2.

d.2. Die Thalesellipsen

d.2.a. *Oberfläche*: Gesucht sei die von der Thalesellipse auf der Kugel begrenzte Fläche. Hier gilt entsprechend Absatz d.1.a. mit der unteren Grenze $\varphi = 0$ und der oberen Grenze $\varphi = \pi/2$ jetzt

$$A = 2R^2\int_0^{\pi/2}(1-\cos\lambda)\,d\varphi\,. \tag{12}$$

Nach Gleichung (1) ist aber

$$\cos\lambda = 1/\sqrt{1+\tan^2\delta\sin^2\varphi}\,, \tag{13a}$$

oder wenn man die Substitution $\psi = \pi/2 - \varphi$ einführt,

$$\cos\lambda = (\cos\delta)/\Delta \tag{13b}$$

mit der Abkürzung

$$\Delta = \sqrt{1-k^2\sin^2\psi} \tag{14a}$$

und

$$k = \sin\delta\,. \tag{14b}$$

Dann folgt

$$A = 2R^2\left(\frac{\pi}{2} + \cos\delta\int_{\pi/2}^{0}\frac{d\psi}{\Delta}\right) = R^2\left[\pi - 2\cos\delta\cdot F\left(k,\frac{\pi}{2}\right)\right] \tag{15}$$

mit F als dem vollständigen elliptischen Integral erster Gattung wobei üblicherweise $k = \sin\alpha$ gesetzt wird, also hier δ gleich α ist (vgl. a. [3]).

d.2.b. *Projektion auf die x,y-Ebene*: Es gilt entsprechend Absatz d.1.b. hier

$$A = 2\cdot\tfrac{1}{2}\int_0^{\pi/2}R^2\sin^2\lambda\,d\varphi = R^2\int_0^{\pi/2}(1-\cos^2\lambda)\,d\varphi =$$

$$= R^2\pi/2 - R^2A_2\,. \tag{16}$$

Drückt man den Integranden $(1+\tan^2\delta\sin^2\varphi)^{-1}$ in A_2 durch $\tan\varphi$ aus und substituiert $\tan\varphi = \mu\cos\delta$, so wird

$$A_2 = \cos\delta\cdot\operatorname{arc}\tan\left(\frac{\tan\varphi}{\cos\delta}\right)\Bigg|_0^{\pi/2} = \frac{\pi}{2}\cos\delta\,.$$

Geht man hiermit in Gleichung (16) ein, so erhält man für die gesamte Fläche

$$A = \tfrac{1}{2}R^2\pi(1-\cos\delta) = \left(R\sin\frac{\delta}{2}\right)^2\pi\,. \tag{17}$$

Diese Fläche ist somit gleich dem Inhalt des Kreises, in welchem die Kugel durch eine durch M gehende und senkrecht zur z-Achse stehende Ebene geschnitten wird (vgl. Bild 1b), oder auch gleich dem Inhalt des in der Projektion (Bild 1a) eingezeichneten

Kreises vom Radius $\overline{MM_z}$ (Kurve c). Beim Übergang zur ebenen Bewegung wird $A = \pi(d/2)^2$, d. h. trivial gleich dem Inhalt des Kreises vom Durchmesser d.

Literatur

1 Meyer zur Capellen, W., Die sphärischen Analoga zu den ϱ-Kurven von H. Alt. Mechanism and Machine Theory 17 (1982).
2 Dittrich, G., Über die momentane Bewegungsgeometrie eines sphärisch bewegten starren Systems, Diss. TH Aachen 1964.
3 Meyer zur Capellen, W., Integraltafeln, Springer-Verlag Berlin-Göttingen-Heidelberg 1950.

Eingereicht am 20. 11. 1980

Anschrift: o. Prof. em. Dr.-Ing. Walther Meyer zur Capellen, Ronheider Weg 46, D-5100 Aachen, BRD

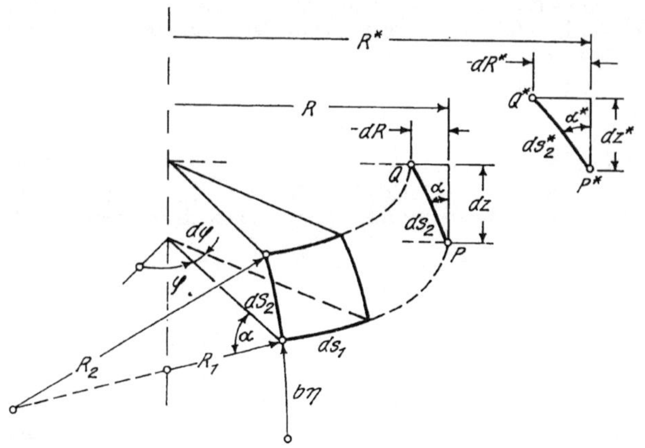

Abb. 1. Verzerrungsgeometrie der deformierten Rotationsschale

ZAMM 62, 404 – 406 (1982)

W. Hübner / F. A. Emmerling

Axialsymmetrische große Deformationen einer elastischen Kegelschale

Die axialsymmetrischen großen Deformationen von Rotationsschalen werden durch zwei nichtlineare Differentialgleichungen vom Reissner-Meissner-Typ beschrieben. Ihre Ableitung wird skizziert und auf die Kegelschale spezifiziert. Die Differentialgleichungen werden mit Hilfe von Fourierreihenansätzen der Lösungsfunktionen integriert. Dies folgt in einem von E. L. Axelrad vorgeschlagenen Matrizenkalkül. Zusätzlich werden die nichtlinearen Differentialgleichungen numerisch integriert.

1. Verzerrungsgeometrie

Die Mittelfläche der Rotationsschale wird durch orthogonale Gausssche Parameter φ, η (Abb. 1) beschrieben. Da eine vollständig (geometrisch) nichtlineare Theorie abgeleitet wird, werden sowohl eine Ausgangskonfiguration als auch eine endlich deformierte Momentankonfiguration betrachtet. Die geometrischen Größen der Momentankonfiguration werden mit einem hochgestellten Stern gekennzeichnet. Es sind $ds_1 = R\,d\varphi$, $ds_2 = b\,d\eta$ die infinitesimalen Bogenelemente auf den Parameterlinien und R_1, R_2 die Krümmungsradien in den Normalschnitten an die Parameterlinien. In der Momentankonfiguration gelten $ds_1^* = (1 + \varepsilon_1)\,ds_1$, $ds_2^* = (1 + \varepsilon_2)\,ds_2$. Aus Abb. 1 entnimmt man folgende Beziehungen

$$\frac{1}{R_1} = \frac{\cos\alpha(\eta)}{R}, \qquad \frac{1}{R_1^*} = \frac{\cos\alpha^*(\eta)}{R^*} = \frac{\cos\alpha^*}{(1+\varepsilon_1)\,R},$$

$$\frac{1}{R_2} = \frac{d\alpha}{ds_2} = \frac{d\alpha}{b\,d\eta}, \qquad \frac{1}{R_2^*} = \frac{d\alpha^*}{ds_2^*} = \frac{d\alpha^*}{(1+\varepsilon_2) b\,d\eta}. \tag{1}$$

Die Krümmungsänderungen (vgl. [1]) ergeben sich zu

$$\varkappa_1 = \frac{1+\varepsilon_1}{R_1^*} - \frac{1}{R_1} = -\frac{\cos\alpha - \cos\alpha^*}{R},$$

$$\varkappa_2 = \frac{1+\varepsilon_2}{R_2^*} - \frac{1}{R_2} = \frac{\dot\alpha^* - \dot\alpha}{b} = \frac{\dot\vartheta}{b}. \tag{2}$$

Hierbei sind $\alpha(\eta)$, $\alpha^*(\eta)$ die Tangentenneigungswinkel zugeordneter Meridiane gegenüber der z-Achse (Abb. 1), $\vartheta = \alpha^* - \alpha$ ist der Tangentendrehwinkel und der übergesetzte Punkt bezeichnet die Ableitung nach η: $()^{\cdot} = d()/d\eta$.

Aus der Definitionsgleichung für die Dehnung in φ-Richtung

$$\varepsilon_1 = \frac{ds_1^* - ds_1}{ds_1} = \frac{R^* - R}{R}$$

wird nach Differentiation bezüglich η unter Beachtung von

$$-dR = \sin\alpha\, b d\eta, \qquad -dR^* = (1+\varepsilon_2)\sin\alpha^*\, b\, d\eta$$

eine nichtlineare Kompatibilitätsbedingung

$$(\varepsilon_1 R)^{\cdot} + b(1+\varepsilon_2)\sin\alpha^* - b\sin\alpha = 0 \tag{3}$$

gewonnen. Weiterhin gilt für die Verschiebung w in z-Richtung

$$dw = dz^* - dz = (1+\varepsilon_2)\cos\alpha^*\, b\, d\eta - \cos\alpha\, b\, d\eta,$$

bzw.

$$\dot w = -b[\cos\alpha - (1+\varepsilon_2)\cos\alpha^*]. \tag{4}$$

2. Schnittlasten, Gleichgewichtsbedingungen, Materialgleichungen

Das Gleichgewicht der Kräfte und Momente wird in der Momentankonfiguration aufgestellt. Da angenommen wird, daß Spannungs- und Deformationszustand rotationssymmetrisch sind, verschwinden die Schubkräfte, Torsionsmomente und die Querkraft im Schnitt φ = const.

Es ist zweckmäßig (Abb. 2), die Schnittlasten T_R und T_Z im Schnitt η = const. in Richtung der Koordinatenachsen einzuführen. Mit diesen erhält man

$$T_2 = -T_R\sin\alpha^* + T_Z\cos\alpha^*, \quad Q_2 = T_R\cos\alpha^* + T_Z\sin\alpha^*. \tag{5}$$

Die Kräftegleichgewichtsbedingungen ohne Flächenlasten lauten

$$(T_R R)^{\cdot} - T_1 b = 0, \qquad (T_Z R)^{\cdot} = 0. \tag{6), (7}$$

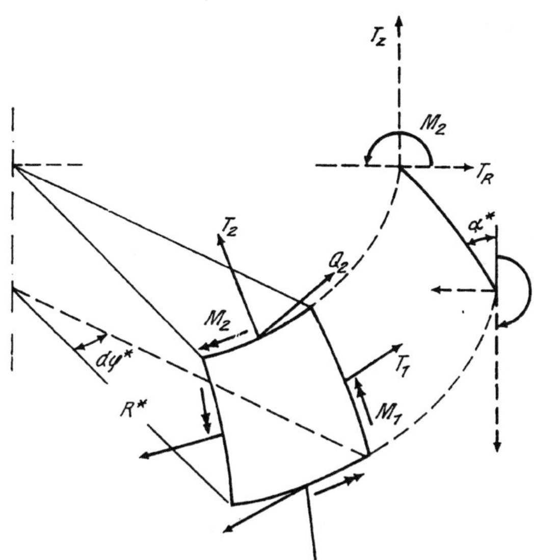

Abb. 2. Schnittlasten in der Momentankonfiguration

Mit der Annahme kleiner Verzerrungen ($\varepsilon_2 \ll 1$) erhält man die Momentengleichgewichtsbedingung zu

$$(M_2 R)^{\cdot} + M_1 b\sin\alpha^* - T_R R b\cos\alpha^* - T_Z R b\sin\alpha^* = 0. \tag{8}$$

Es wird rein elastisches Materialverhalten vorausgesetzt, so daß man mit dem Elastizitätsmodul E, der Querkontraktionszahl ν, der Schalendicke h und der Biegesteifigkeit der Schale $D = Eh^3/12(1-\nu^2)$ die Materialgleichungen

$$Eh\varepsilon_1 = T_1 - \nu T_2\,, \qquad Eh\varepsilon_2 = T_2 - \nu T_1\,, \tag{9}$$

$$M_1 = D(\varkappa_1 + \nu\varkappa_2)\,, \qquad M_2 = D(\varkappa_2 + \nu\varkappa_1)\,. \tag{10}$$

erhält.

3. Reissner-Meissner-Gleichungen

Die vorliegenden Gleichungen können auf zwei nichtlineare Differentialgleichungen vom Reissner-Meissner-Typ reduziert werden. Um diese zu erhalten, werden zuerst in (9) die Normalkräfte T_1, T_2 durch die Schnittlasten T_R, T_Z nach (5) und (6) ersetzt und die Krümmungsänderungen $\varkappa_1$, $\varkappa_2$ in (10) mittels (2) eliminiert:

$$Eh\varepsilon_1 = \frac{1}{b}(T_R R)^{\boldsymbol\cdot} + \nu(T_R \sin\alpha^* - T_Z \cos\alpha^*)\,,$$

$$Eh\varepsilon_2 = -T_R \sin\alpha^* + T_Z \cos\alpha^* - \frac{\nu}{b}(T_R R)^{\boldsymbol\cdot}\,, \tag{11}$$

$$bM_1 = \frac{D}{R}[-b(\cos\alpha - \cos\alpha^*) + \nu R\dot\vartheta],$$

$$RM_2 = \frac{D}{b}[R\dot\vartheta - \nu b(\cos\alpha - \cos\alpha^*)]\,. \tag{12}$$

Die erste Lösungsgleichung folgt aus der Kompatibilitätsbedingung (3), in der die Dehnungen mit (11) ersetzt werden,

$$[(T_R R)^{\boldsymbol\cdot} R]^{\boldsymbol\cdot} - \left[\frac{b^2}{R^2}\sin^2\alpha^* - \nu\frac{b}{R}(\sin\alpha^*)^{\boldsymbol\cdot}\right](T_R R)\,R\, +$$

$$+\, Ehb^2(\sin\alpha^* - \sin\alpha) = [-\nu bR(\cos\alpha^*)^{\boldsymbol\cdot} - b^2\cos\alpha^*\sin\alpha^*]\,T_Z\,. \tag{13}$$

Durch Einsetzen der Biegemomente (12) in die Momentengleichgewichtsbedingung (8) erhält man die zweite Lösungsgleichung zu

$$(R\dot\vartheta)^{\boldsymbol\cdot} - \left[\frac{b^2}{R^2}(\cos\alpha - \cos\alpha^*)\sin\alpha^* + \nu\frac{b}{R}(\cos\alpha - \cos\alpha^*)^{\boldsymbol\cdot}\right]R\, +$$

$$+\, \nu b\dot\vartheta\sin\alpha^* = \frac{b^2}{D}(T_R R\cos\alpha^* + T_Z R\sin\alpha^*)\,. \tag{14}$$

In den beiden gekoppelten nichtlinearen Differentialgleichungen (13, 14) sind die unterstrichelten Terme gegenüber den anderen Termen in den Gleichungen klein von höherer Ordnung und bleiben unberücksichtigt. Eine mathematische Begründung findet man in [1].

4. Randbedingungen und Normierung

Mit den vorliegenden Gleichungen wird speziell die Deformation einer Kegelschale (Abb. 3) unter einer Kraft P_3 berechnet. Der obere Rand der Schale kann sich frei verschieben. Am unteren Rand liegt die Kegelschale in radialer Richtung frei verschieblich auf.

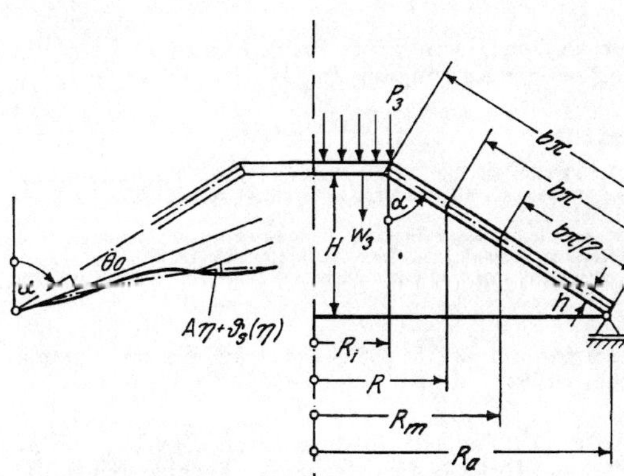

Abb. 3. Tellerfeder mit Maßbezeichnungen und deformierter Meridianlinie

Damit gilt wegen (7) im gesamten Bereich

$$T_Z R = \text{const.} = -\frac{P_3}{2\pi} = \frac{P}{2\pi}\,. \tag{15}$$

Die Randbedingungen bei $R = R_i$ und $R = R_a$ für den vorliegenden Fall lauten

$$T_R R = 0 \tag{16}$$

und wegen $M_2 = 0$

$$R\dot\vartheta - \nu b(\cos\alpha - \cos\alpha^*) = 0\,. \tag{17}$$

Es ist zweckmäßig, sowohl die Differentialgleichungen (13), (14) als auch die Randbedingungen (16), (17) durch Einführung von dimensionslosen Größen zu vereinfachen.
Mit

$$r = \frac{R}{R_m}\,; \quad \chi = \frac{b}{R_m}\,; \quad h^0 = \frac{h}{b\sqrt{12(1-\nu^2)}}\,; \quad \mu = \frac{\chi}{h^0}\,,$$

$$\psi = T_R R\frac{h^0 b}{D}\,; \quad P_3^0 = \frac{P_3}{2\pi}\frac{b^2}{R_m D} \tag{18}$$

gehen die reduzierten Gleichungen (13), (14) in

$$(r\dot\psi)^{\boldsymbol\cdot} - \frac{\chi^2}{r}\sin^2\alpha^*\psi + \mu(\sin\alpha^* - \sin\alpha) = 0\,, \tag{19}$$

$$(r\dot\vartheta)^{\boldsymbol\cdot} - \frac{\chi^2}{r}(\cos\alpha - \cos\alpha^*)\sin\alpha^* - \mu\psi\cos\alpha^* = -P_3^0\sin\alpha^* \tag{20}$$

über, während die Randbedingungen (16), (17) damit

$$\psi = 0\,, \qquad r\dot\vartheta - \nu\chi(\cos\alpha - \cos\alpha^*) = 0 \tag{21), (22}$$

lauten.

Die in (18) eingeführte Größe ψ kann wegen (6) als Spannungsfunktion interpretiert werden. Für den dimensionslosen Radius $r(\eta)$ gilt nach Abb. 3 und (18)

$$r = 1 + \left(\frac{\pi}{2} - \eta\right)\chi\sin\alpha\,. \tag{23}$$

5. Integration der Differentialgleichungen

Die Integration der nichtlinearen Differantialgleichungen (19) und (20) wurde mit 2 verschiedenen Methoden durchgeführt.

Bei der ersten werden (19) und (20) mit Hilfe des Runge-Kutta-Verfahrens integriert. Dazu werden diese Differentialgleichungen in ein System von Differentialgleichungen 1. Ordnung umgeformt, wobei die Zustandsgrößen ϑ, $r\dot\vartheta$, ψ, $r\dot\psi$ verwendet werden und $r(\eta)$ mit (23) gegeben ist. Die 2 Randbedingungen (21) und (22) gelten bei $\eta = 0$ und $\eta = \pi$. Die Integration muß deshalb mit zwei unbekannten Anfangswerten $\vartheta(0)$ und $r\dot\psi(0)$ durchgeführt werden. Letztere werden dabei solange variiert, bis (21) und (22) bei $\eta = \pi$ erfüllt sind. Obwohl hierfür eine gute Suchstrategie entwickelt wurde, ist diese Methode für die Berechnung von Federkennlinien, wie sie hier interessieren, sehr aufwendig.

Durch das im folgenden beschriebene Verfahren gelingt die Integration ohne Suchstrategie mit wenigen Iterationen. Dazu werden die Funktionen $\vartheta(\eta)$ und $\psi(\eta)$ in Fourierreihen entwickelt, deren Koeffizienten als Spaltenmatrizen geschrieben und der von E. L. Axelrad vorgeschlagene, in [1] und [3] ausführlicher beschriebene Matrizen-Kalkül angewandt. Für die Anwendung im vorliegenden Fall müssen die Multiplikation und die Differentiation von Fourierreihen beschrieben werden:
Die Funktion

$$a(\eta) = \frac{a_{c0}}{2} + a_{c1}\cos\eta + a_{c2}\cos 2\eta + \ldots + a_{s1}\sin\eta\, +$$

$$+\, a_{s2}\sin 2\eta + \ldots \tag{24}$$

wird durch die Spaltenmatrizen $\boldsymbol{a}_c$ und $\boldsymbol{a}_s$ der Fourierkoeffizienten repräsentiert, wobei gilt

$$\boldsymbol{a}_c = \left[\frac{a_{c0}}{2},\ a_{c1},\ a_{c2}, \ldots\right]^T; \quad \boldsymbol{a}_s = [a_{s1},\ a_{s2}, \ldots]^T\,. \tag{25}$$

Das Produkt $p(\eta) = a(\eta)\,b(\eta)$ wird durch die Matrizengleichung

$$p = ab \div \begin{bmatrix}\boldsymbol{p}_c\\ \boldsymbol{p}_s\end{bmatrix} = \begin{bmatrix}\boldsymbol{a}_{c+} & \boldsymbol{a}_{s-}\\ \boldsymbol{a}_{s+} & \boldsymbol{a}_{c-}\end{bmatrix}\begin{bmatrix}\boldsymbol{b}_c\\ \boldsymbol{b}_s\end{bmatrix} \tag{26}$$

ersetzt, wobei $\boldsymbol{b}_c$, $\boldsymbol{p}_c$ und $\boldsymbol{b}_s$, $\boldsymbol{p}_s$ wie $\boldsymbol{a}_c$ bzw. $\boldsymbol{a}_s$ aufgebaut sind. Die Differentiation von (24) ergibt

$$a(\eta) = a_{s1} \cos\eta + 2a_{s2} \cos 2\eta + \ldots - a_{c1} \sin\eta - 2a_{c2} \sin 2\eta - \ldots,$$

in Matrizenform also

$$\dot{q} = \dot{a} \mathrel{\hat{=}} \begin{bmatrix} \boldsymbol{q}_c \\ \boldsymbol{q}_s \end{bmatrix} = \begin{bmatrix} \boldsymbol{0} & \boldsymbol{A}_- \\ -\boldsymbol{A}_+ & \boldsymbol{0} \end{bmatrix} \begin{bmatrix} \boldsymbol{a}_c \\ \boldsymbol{a}_s \end{bmatrix}. \tag{27}$$

Da im allgemeinen die Funktionen entweder als Sinus- oder Cosinusfunktion angesetzt werden, benötigt man die 6 aus (26) und (28) abzulesenden Sonderfälle, wobei die Plus- und Minusmatrizen mit entsprechenden Unterprogrammen berechnet werden.

Mit Hilfe der formalisierten Multiplikation lassen sich auch beliebige Funktionen darstellen. So schreibt man zum Beispiel

$$\cos\alpha^* = \cos(\alpha + \eta) = c_0^*/2 + c_1^* \cos\eta + c_2^* \cos 2\eta + \ldots, \tag{28}$$

berechnet $c_0^*, c_1^*, c_2^*, \ldots$ mit der Reihenentwicklung und benützt im folgenden die Entsprechungen

$$\cos\alpha^* \mathrel{\hat{=}} \boldsymbol{c}^* = [c_0^*/2,\; c_1^*,\; c_2^*,\; \ldots]^T,$$

$$\sin\alpha^* \mathrel{\hat{=}} \boldsymbol{s}^* = [s_1^*,\; s_2^*,\; \ldots]^T. \tag{29}$$

Bei dem vorliegenden Problem werden die Funktionen $e_c = 1$, $e_s = 1$ und $\eta_s = \eta$ für $0 \leqq \eta \leqq \pi$ benötigt und durch $\boldsymbol{e}_c$, $\boldsymbol{e}_s$ bzw. $\boldsymbol{\eta}_s$ repräsentiert.
Außerdem erweist es sich als sinnvoll, die Funktionen c' und s' mit

$$\sin\alpha^* - \sin\alpha = c'\vartheta\,; \qquad \cos\alpha - \cos\alpha^* = s'\vartheta \tag{30}$$

einzuführen. Für die Variablen der Differentialgleichung werden nun die Ansätze

$$\vartheta(\eta) = \Theta_0 + A\eta + \vartheta_s \quad \text{mit} \quad \vartheta_s(\eta) = \vartheta_{s1} \sin\eta + \vartheta_{s2} \sin 2\eta + \ldots, \tag{31}$$

$$\psi(\eta) = \psi_1 \sin\eta + \psi_2 \sin 2\eta + \psi_3 \sin 3\eta + \ldots, \tag{32}$$

$$r(\eta) = \frac{r_0}{2} + r_1 \cos\eta + r_2 \cos 2\eta + \ldots \tag{33}$$

mit den entsprechenden Spaltenmatrizen $\boldsymbol{\vartheta}_s$, $\boldsymbol{\psi}$, $\boldsymbol{r}$ gemacht.

Der Ansatz (31) erweist sich wegen des großen konstanten Anteils Θ_0 als sinnvoll; da außerdem $\vartheta_s(\eta)$ an den Rändern verschwindet, muß $A(\eta)$ hinzugenommen werden (vgl. Abb. 3). Mit (31) und (23) erhält man

$$(r\dot{\vartheta})^{\cdot} = -A\chi \sin\alpha + (r\dot{\vartheta}_s)^{\cdot}, \tag{34}$$

so daß man nun anstelle der Differentialgleichungen (19) und (20) die beiden Matrizengleichungen

$$-\boldsymbol{A}_+\boldsymbol{r}_+\boldsymbol{A}_-\boldsymbol{\psi} - \chi^2\boldsymbol{r}_-^{-1}\boldsymbol{s}_+^*\boldsymbol{s}_-^*\boldsymbol{\psi} + \mu\boldsymbol{c}'_-(\Theta_0\boldsymbol{e}_s + A\boldsymbol{\eta}_s + \boldsymbol{\vartheta}_s) = \boldsymbol{0}, \tag{35}$$

$$-\boldsymbol{A}_+\boldsymbol{r}_+\boldsymbol{A}_-\boldsymbol{\vartheta}_s - \chi A \sin\alpha\, \boldsymbol{e}_s - \chi^2\boldsymbol{r}_-^{-1}\boldsymbol{s}_+^*\boldsymbol{s}'_-(\Theta_0\boldsymbol{e}_s + A\boldsymbol{\eta}_s + \boldsymbol{\vartheta}_s) -$$
$$- \mu\boldsymbol{c}_-^*\boldsymbol{\psi} = -P_3^0\boldsymbol{s}^* \quad \text{erhält.} \tag{36}$$

Aus (35) wird $\boldsymbol{\psi}$ berechnet und in (36) eingesetzt. Wie man leicht erkennt, erhält man daraus die Linearkombination

$$\boldsymbol{\vartheta}_s = \Theta_0\boldsymbol{d}_1 + A\boldsymbol{d}_2 + P_3^0\boldsymbol{d}_3 \tag{37}$$

mit den Spaltenmatrizen $\boldsymbol{d}_1$, $\boldsymbol{d}_2$ und $\boldsymbol{d}_3$, die aus den in (35) und (36) vorkommenden Matrizen gebildet werden.

Die Randbedingung (21) ist durch den Ansatz (32) bereits erfüllt, während aus (22) mit (30) folgt

$$\dot{\vartheta} - \frac{\nu}{r}\chi s'\vartheta = 0 \quad \text{für } r = r_i = R_i/R_m \text{ und } r = r_a = R_a/R_m. \tag{38}$$

Aus (31) folgt durch Differentiation

$$\dot{\vartheta} = A + \vartheta_{s1} \cos\eta + \vartheta_{s2}\, 2 \cos 2\eta + \vartheta_{s3}\, 3 \cos 3\eta + \ldots. \tag{39}$$

Mit

$$\boldsymbol{F}_0^T = [1;\ 2;\ 3;\ \ldots] \quad \text{und} \quad \boldsymbol{F}_\pi^T = [-1;\ +2;\ -3;\ -+\ldots] \tag{40}$$

folgen also aus (38) die 2 skalaren Gleichungen.

$$A + \boldsymbol{F}_0^T\boldsymbol{\vartheta}_s - \frac{\nu}{r_a}\chi s'\Theta_0 = 0\,, \qquad A + \boldsymbol{F}_\pi^T\boldsymbol{\vartheta}_s - \frac{\nu}{r_i}\chi s'(\Theta_0 +$$
$$+ A\pi) = 0 \tag{41}$$

in der Form

$$\begin{bmatrix} a_{11} & a_{12} \\ a_{21} & a_{22} \end{bmatrix} \begin{bmatrix} \Theta_0 \\ A \end{bmatrix} = -\begin{bmatrix} a_{13} \\ a_{23} \end{bmatrix} P_3^0. \tag{42}$$

Da die Koeffizienten a_{ij} von ϑ abhängen, müssen die Gleichungen (42) iterativ gelöst werden. Dabei ist es numerisch günstiger Θ_0 vorzugeben und P_3^0 sowie A zu berechnen. Zur Darstellung der Federkennlinie wird die Verschiebung w_3 des Lastangriffspunktes benötigt, die man aus (4) bei $\varepsilon_2 \ll 1$ und (28) zu

$$w_3 = -\int_0^\pi \dot{w}\,d\eta = b\pi\left(\cos\alpha - \frac{c_0^*}{2}\right). \tag{43}$$

erhält.

6. Ergebnisse

Mit der beschriebenen Methode lassen sich nun auf bequeme Weise Federkennlinien $P_3(w_3)$ berechnen, wie sie für die technische Anwendung wichtig sind. Abb. 4 zeigt am Beispiel einer Tellerfeder den typischen Verlauf. Zum Vergleich ist die bisher verwendete, nach [2] berechnete Kennlinie gestrichelt eingezeichnet. Die Ergebnisse der hier verwendeten exakten nichtlinearen Theorie zeigen, daß die Tellerfedern weicher sind, als die bisherigen Berechnungen nach [2] ergaben. Eine ausführliche vergleichende Parameterstudie einschließlich der Spannungsberechnung ist in [4] zu finden.

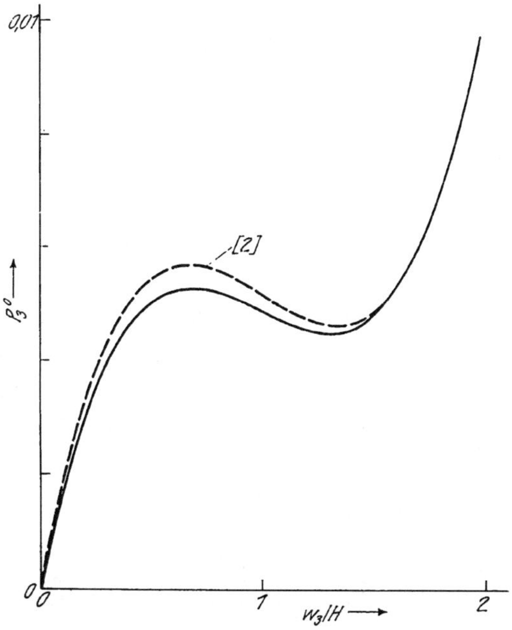

Abb. 4. Last-Durchbiegungskurve einer Tellerfeder im Vergleich zu den Ergebnissen von ALMEN/LASZLO ([2])

Die Verfasser danken Herrn Prof. Dr. E. L. AXELRAD für die Unterstützung bei der Anfertigung der Arbeit.

Literatur

1 AXELRAD, E. L., Flexible Schalen, Verlag Nauka, Moskau 1976. (in Russisch).
2 ALMEN, J. O.; LASZLO, A., The uniform-section disk spring, Trans. ASME, vol. 58 (1936), 305–314
3 EMMERLING, F. A., Nichtlineare Biegung und Beulen von Zylindern und krummen Rohren bei Normaldruck. Ingenieurarchiv, 52 (1982).
4 HÜBNER, W., Deformationen und Spannungen bei Tellerfedern. Konstruktion 34 (1982).

Eingereicht am 21. 5. 1981 (Vortrag GAMM-Tagung Würzburg 1981)

Anschrift: Dr.-Ing. W. HÜBNER/Prof. Dr.-Ing. F. A. EMMERLING, Institut für Mechanik, Fachbereich Luft- und Raumfahrttechnik, Hochschule der Bundeswehr München, Werner-Heisenberg-Weg 39, 8014 Neubiberg, BRD

ZAMM 62, 407 – 408 (1982)

R. N. DUBEY / A. MEYERS

Grafische Darstellung der Streckungen bei ebener Verzerrung

Bei der Bestimmung der Verzerrungstensoren fällt den Streckungstensoren eine besondere Rolle zu. Anhand einer grafischen Darstellung kann der Zusammenhang zwischen den beiden Streckungstensoren und dem Deformationsgradienten anschaulich gedeutet werden.

Bezeichnet $\boldsymbol{X}$ den Ortsvektor eines materiellen Punktes in der Referenzkonfiguration, $\boldsymbol{x}$ den Ortsvektor zum gleichen materiellen Punkt in der aktuellen Konfiguration, so ist der Deformationsgradient in folgender Weise definiert:

$$\boldsymbol{F} = \frac{\partial \boldsymbol{x}}{\partial \boldsymbol{X}} . \tag{1}$$

Der Deformationsgradient $\boldsymbol{F}$ läßt sich multiplikativ zerlegen in den orthogonalen Rotationstensor $\boldsymbol{R}$ und in den Lagrange-Streckungstensor $\boldsymbol{U}$ beziehungsweise den Eulerstreckungstensor $\boldsymbol{V}$. Bei der Zerlegung

$$\boldsymbol{F} = \boldsymbol{R}\boldsymbol{U} \tag{2}$$

wird der Ausgangszustand über die rechte Streckung $\boldsymbol{U}$ und anschließende Rotation $\boldsymbol{R}$ in den Endzustand übergeführt. Bei der Zerlegung

$$\boldsymbol{F} = \boldsymbol{V}\boldsymbol{R} \tag{3}$$

erhält man den Endzustand über die der Drehung $\boldsymbol{R}$ folgende Streckung $\boldsymbol{V}$. Der konventionelle Weg, um aus dem Deformationsgradienten $\boldsymbol{F}$ die Streckungstensoren $\boldsymbol{U}$ und $\boldsymbol{V}$ sowie den Rotationstensor $\boldsymbol{R}$ zu erhalten, geht über die Berechnung der Quadrate der Streckungstensoren:

$$\boldsymbol{U}^2 = \boldsymbol{F}^T\boldsymbol{F} , \qquad \boldsymbol{V}^2 = \boldsymbol{F}\boldsymbol{F}^T . \tag{4}$$

Bei Beachtung der Symmetrie der Streckungstensoren läßt sich hieraus $\boldsymbol{U}$ und $\boldsymbol{V}$ ermitteln. Den Rotationstensor erhält man anschließend aus dem Produkt des Deformationsgradienten mit der Inversen der Streckung:

$$\boldsymbol{R} = \boldsymbol{F}\boldsymbol{U}^{-1} = \boldsymbol{V}^{-1}\boldsymbol{F} . \tag{5}$$

Im ebenen Verzerrungsfall kann man nachfolgend beschriebenen Weg einschlagen.

Der Rotation eines materiellen Linienelementes mit dem Winkel θ im Gegenuhrzeigersinn entspricht der Rotationstensor:

$$\boldsymbol{R}(\theta) = \begin{pmatrix} \cos\theta & -\sin\theta \\ \sin\theta & \cos\theta \end{pmatrix} . \tag{6}$$

Um spätere Zusammenhänge einfacher darstellen zu können, wird ein zweiter Rotationstensor eingeführt. Er entspricht der oben angeführten Rotation mit einer zusätzlichen spiegelbildlichen Abbildung um die 1-Achse:

$$\boldsymbol{R}^*(\Theta) = \begin{pmatrix} \cos\Theta & \sin\Theta \\ \sin\Theta & -\cos\Theta \end{pmatrix} . \tag{7}$$

Folgende Beziehungen gelten:

$$\boldsymbol{R}(\theta)\,\boldsymbol{R}^*(\Theta) = \boldsymbol{R}^*(\theta + \Theta) , \qquad \boldsymbol{R}^T(\theta)\,\boldsymbol{R}^*(\Theta) = \boldsymbol{R}^*(\theta - \Theta) . \tag{8}$$

Ist $\boldsymbol{R}$ bekannt, so läßt sich die Streckung beschreiben zu

$$\boldsymbol{U} = \boldsymbol{R}^T\boldsymbol{F} , \qquad \boldsymbol{V} = \boldsymbol{F}\boldsymbol{R}^T . \tag{9}$$

Insbesondere gilt

$$U_{21} = F_{12}\cos\theta + F_{22}\sin\theta , \qquad U_{12} = -F_{11}\sin\theta + F_{21}\cos\theta . \tag{10}$$

Bei Beachtung der Symmetrie von $\boldsymbol{U}$ folgt:

$$\tan\theta = \frac{F_{21} - F_{12}}{F_{11} + F_{22}} . \tag{11}$$

Es werden nun drei neue Größen s, t, α eingeführt, welche folgendermaßen definiert sind:

$$F_{11} + F_{22} = 2s\cos\theta , \qquad F_{11} - F_{22} = 2t\cos\alpha ,$$
$$F_{21} - F_{12} = 2s\sin\theta , \qquad F_{21} + F_{12} = 2t\sin\alpha , \tag{12}$$

beziehungsweise

$$2s = \sqrt{(F_{11} + F_{22})^2 + (F_{21} - F_{12})^2} ,$$
$$2t = \sqrt{(F_{11} - F_{22})^2 + (F_{21} + F_{12})^2} , \tag{13}$$
$$\tan\theta = \frac{F_{21} - F_{12}}{F_{11} + F_{22}} , \qquad \tan\alpha = \frac{F_{21} + F_{12}}{F_{11} - F_{22}} .$$

$\boldsymbol{F}$ beschreibt sich mit diesen Größen zu

$$\boldsymbol{F} = s\boldsymbol{R}(\theta) + t\boldsymbol{R}^*(\alpha) . \tag{14}$$

Die Streckungstensoren ergeben sich zu

$$\boldsymbol{U} = \boldsymbol{R}^T(\theta)\,\boldsymbol{F} = s\,\boldsymbol{I} + t\boldsymbol{R}^*(2\beta_L) , \tag{15}$$

$$\boldsymbol{V} = \boldsymbol{F}\boldsymbol{R}^T(\theta) = s\boldsymbol{I} + t\boldsymbol{R}^*(2\beta_E) \tag{16}$$

mit

$\boldsymbol{I} \mathrel{\hat{=}}$ Einheitstensor,

$\beta_L = (\alpha - \theta)/2$, $\quad \beta_E = (\alpha + \theta)/2$.

Setzt man

$$\bar{\boldsymbol{U}} = \begin{pmatrix} (U_{11} - U_{22})/2 & U_{12} \\ U_{12} & -(U_{11} - U_{22})/2 \end{pmatrix} ,$$
$$\bar{\boldsymbol{V}} = \begin{pmatrix} (V_{11} - V_{22})/2 & V_{12} \\ V_{12} & -(V_{11} - V_{22})/2 \end{pmatrix} , \tag{17}$$

so lassen sich die Zusammenhänge zwischen $\boldsymbol{U}$ und $\boldsymbol{V}$ auch angeben zu

$$\boldsymbol{U} = \boldsymbol{R}^T\boldsymbol{V}\boldsymbol{R} = s\boldsymbol{I} + \bar{\boldsymbol{V}}\boldsymbol{R}(2\theta) , \tag{18}$$

$$\boldsymbol{V} = \boldsymbol{R}\boldsymbol{U}\boldsymbol{R}^T = s\boldsymbol{I} + \bar{\boldsymbol{U}}\boldsymbol{R}(2\theta) . \tag{19}$$

Damit gilt auch

$$\bar{\boldsymbol{V}}\boldsymbol{R}(2\theta) = t\boldsymbol{R}^*(2\beta_L) , \qquad \bar{\boldsymbol{U}}\boldsymbol{R}(2\theta) = t\boldsymbol{R}^*(2\beta_E) . \tag{20}$$

Die in den Formeln gegebenen Zusammenhänge lassen sich anschaulich anhand eines Kreises („MOHRscher" Kreis für Streckungen) zeigen (siehe Abb.). Auf der Abszissenachse werden die Diagonalelemente der Streckungstensoren, auf der Ordinatenachse die übrigen Elemente aufgetragen. Der Mittelpunkt des Kreises liegt auf der Abszissenachse und ist bestimmt durch

$$s = \frac{1}{2}\sqrt{(F_{11} + F_{22})^2 + (F_{21} - F_{12})^2} =$$
$$= (U_{11} + U_{22})/2 = (V_{11} + V_{22})/2 , \tag{21}$$

wie aus (13), (15) und (16) hervorgeht.

Aus (13), (18) und (19) erhält man für den Radius:

$$t = \frac{1}{2}\sqrt{(F_{11} - F_{22})^2 + (F_{21} + F_{12})^2}$$
$$= \frac{1}{2}\sqrt{(U_{11} - U_{22})^2 + 4U_{12}^2} = \frac{1}{2}\sqrt{(V_{11} - V_{22})^2 + 4V_{12}^2} . \tag{22}$$

Die Hauptstreckungen lassen sich sofort ablesen:

$$\lambda_{1,2} = s \pm t . \tag{23}$$

Die Hauptstreckungsrichtungen lassen sich ausgehend von der Richtung eines Linienelementes über die Winkel β_E beziehungsweise β_L bestimmen, je nachdem ob man von der linken Streckung $\boldsymbol{V}$ oder der rechten Streckung $\boldsymbol{U}$ ausgeht (Gleichungen (15) und (16)). Wie man (18) und (19) entnehmen kann, sind die Zeiger für $\boldsymbol{U}$ und $\boldsymbol{V}$ um den Winkel 2θ voneinander entfernt. α, ausgehend von der 1. Hauptstreckungsrichtung, halbiert den Winkel in zwei gleichgroße Teile θ.

Dem Kreis lassen sich einfache Zusammenhänge schnell entnehmen. Bei reiner Streckung ($\theta = 0$) ebenso wie bei einer Rotation um $\theta = n\pi$ sind $\boldsymbol{U}$ und $\boldsymbol{V}$ identisch. Aus (11) entnimmt man, daß der Deformationsgradient in diesem Fall symmetrisch ist. Ist $F_{21} = -F_{12}$, so folgt aus (13,4), daß $\alpha = n\pi$. In diesem Fall sind die Diagonalelemente der beiden Streckungstensoren gleich groß ($U_{11} = V_{11}$, $U_{22} = V_{22}$), während hingegen $U_{12} = -V_{12}$ gilt. Da die zweite Hauptstreckung größer 0 ist, muß der Kreis rechts von der Ordinatenachse liegen. Dies folgt auch nach Einsetzen von (21) und (22) in (23), da die Determinante von $\boldsymbol{F}$ positiv ist.

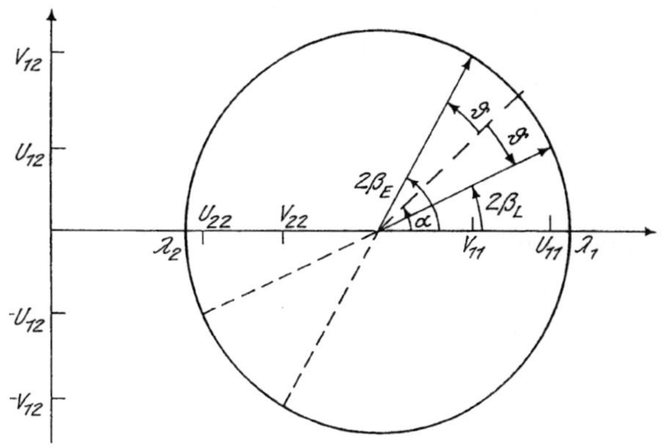

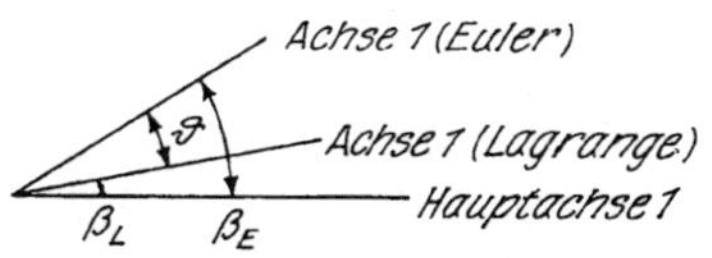

Eingereicht am 19. 6. 1981

Anschrift: Dr.-Ing. ALBERT MEYERS, Institut für Mechanik, Ruhruniversität Bochum, Postfach 2148, D-4630 Bochum, BRD

ZAMM 62, 408–409 (1982)

GY. DÉRI

Berechnung der Grundfrequenz des beiderseits gelenkig gelagerten, biegeschwingenden Stabes mit konstantem Querschnitt, in der Mitte mit einer konzentrierten Masse belastet

Die Masse m_s des in der Abbildung dargestellten Stabes sei über die Länge l gleichmäßig verteilt. Der Stab ist an beiden Enden gelenkig gelagert, die Biegesteifigkeit EI des Querschnittes sei konstant. Der Stab trägt in der Mitte eine Masse m.

Es ist bekannt, daß die kleinste, sogenannte Grundfrequenz, der Biegeschwingungen dieses Stabes

$$\omega_1 = z_1^2 \sqrt{\frac{16EI}{l^3 m_s}} \tag{1}$$

beträgt, wo z_1 die kleinste positive Wurzel der Gleichung

$$\frac{z}{2}(\text{tg}z - \text{th}z) = \frac{m_s}{m} \tag{2}$$

bedeutet [1], die zum Umformen zweckmäßiger in folgender Form geschrieben werden kann:

$$\frac{2z^3}{\text{tg}z - \text{th}z} = \frac{m}{m_s} z^4 . \tag{3}$$

Zur Lösung dieser Gleichung wird die linke Seite durch eine gut approximierende rationale Funktion ausgedrückt. Es würde nicht zum Ziel führen, wenn man die Funktionen tgz und thz in Potenzreihen entwickeln würde, da im Falle $m \to 0$, laut Gl. (2) $z_1 \to \frac{\pi}{2}$ wird, und an der letzteren Stelle beide Potenzreihen divergent sind. Das Ziel wird dadurch erreicht, daß die Funktionen tgz und thz durch allgemeinere, durch ihre Hauptteile ergänzte Potenzreihen ausgedrückt werden [2]:

$$\frac{\text{tg}z}{z} = \frac{2}{\frac{\pi^2}{4} - z^2} + \left(1 - \frac{8}{\pi^2}\right) + \left(\frac{1}{3} - \frac{32}{\pi^4}\right) z^2 + \left(\frac{2}{15} - \frac{128}{\pi^6}\right) z^4 + \left(\frac{17}{315} - \frac{512}{\pi^8}\right) z^6 + \left(\frac{62}{2835} - \frac{2048}{\pi^{10}}\right) z^8 + \left(\frac{1382}{155925} - \frac{8192}{\pi^{12}}\right) z^{10} + \dots, \ |z| < \frac{3}{2}\pi ,$$

$$\frac{\text{th}z}{z} = \frac{2}{\frac{\pi^2}{4} + z^2} + \left(1 - \frac{8}{\pi^2}\right) - \left(\frac{1}{3} - \frac{32}{\pi^4}\right) z^2 + \left(\frac{2}{15} - \frac{128}{\pi^6}\right) z^4 - \left(\frac{17}{315} - \frac{512}{\pi^8}\right) z^6 + \left(\frac{62}{2835} - \frac{2048}{\pi^{10}}\right) z^8 - \left(\frac{1382}{155925} - \frac{8192}{\pi^{12}}\right) z^{10} + \dots, \ |z| < \frac{3}{2}\pi .$$

Von der oberen Gleichung wird die untere subtrahiert, und durch $2z^2$ dividiert; so ergibt sich

$$\frac{\text{tg}z - \text{th}z}{2z^3} = \frac{2}{\frac{\pi^4}{16} - z^4} + \left(\frac{1}{3} - \frac{32}{\pi^4}\right) + \left(\frac{17}{315} - \frac{512}{\pi^8}\right) z^4 + \left(\frac{1382}{155925} - \frac{8192}{\pi^{12}}\right) z^8 + \dots, \ |z| < \frac{3}{2}\pi .$$

Zum Reziprokwert übergehend erhält man nach einiger Rechnung

$$\frac{2z^3}{\text{tg}z - \text{th}z} = 3 - \frac{17}{35} z^4 - \left(\frac{768}{\pi^8} - \frac{1382}{17325}\right) z^8 - \dots, \ |z| < \frac{3}{2}\pi . \tag{4}$$

Der Koeffizient von z^8 beträgt $-0{,}00117$. Dieses und die folgenden Glieder werden vernachlässigt, es ist also

$$\frac{2z^3}{\text{tg}z - \text{th}z} = 3 - 0{,}486\, z^4 . \tag{5}$$

Zur Veranschaulichung der guten Annäherung wurden die Funktionswerte im Bereiche $0 \leqq z \leqq \frac{\pi}{2}$ in einer Tabelle zusammengefaßt:

z	0	0,5	1,0	1,2	1,4	$\pi/2$
$\frac{2z^3}{\text{tg}\,z - \text{th}\,z}$	3	2,970	2,513	1,988	1,117	0
$3 - 0{,}486z^4$	3	2,970	2,514	1,992	1,133	0,041

Aus den Gleichungen (3) und (5) folgt der Wert

$$z^4 = \frac{3}{\frac{m}{m_s} + 0{,}486} ,$$

der, in die Gleichung (1) eingesetzt, den Wert der Grundfrequenz

$$\omega_1 = \sqrt{\frac{48EI}{l^3(m + 0{,}486\, m_s)}} \tag{6}$$

ergibt. Falls in Gl. (4) auch das Glied z^8 in Rechnung gezogen wird, so ergibt sich in den Klammern des Nenners noch die ganz unbedeutende Korrektur $+ \frac{0{,}007\, m_s}{1 + \frac{2m}{m_s}}$.

Ein ähnliches Resultat erreichte T. v. KÁRMÁN für den Fall, daß die Masse m_s des Stabes relativ klein ist [1]. Die unnötige Einschränkung ergab sich dadurch, daß v. KÁRMÁN seine Berechnung mit Potenzreihen durchgeführt hat, deren Konvergenz stark abnimmt, wenn m sich der Null nähert.

Demgegenüber gibt die mitgeteilte Formel auch bei $m = 0$ eine gute Annäherung. In diesem Fall wird nämlich

$$\omega_1 = 9{,}94 \sqrt{\frac{EI}{l^3 m_s}} ,$$

gegenüber dem genauen Wert

$$\omega_1 = \pi^2 \sqrt{\frac{EI}{l^3 m_s}} \approx 9{,}87 \sqrt{\frac{EI}{l^3 m_s}} .$$

Es ist bekannt, daß die Gleichung (6) die biegekritische Drehzahl der mit dem behandelten Stab übereinstimmenden Welle ergibt.

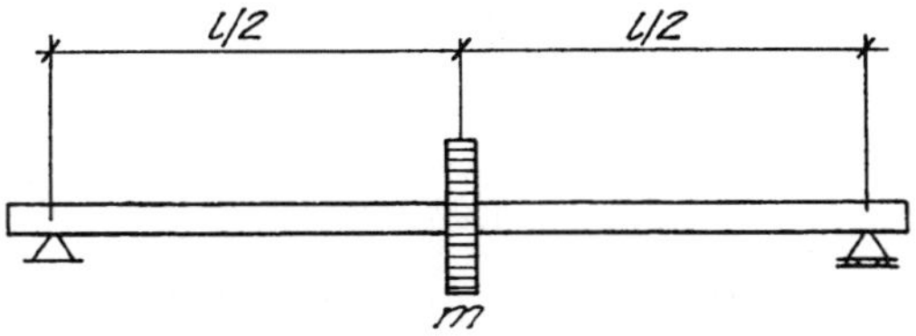

Literatur

1 THEODORE V. KÁRMÁN; MAURICE A. BIOT, Mathematical Methods in Engineering, Kapitel 7 Gl. (95) bzw Gl. (103).
2 JULIUS DÉRI, Die Vergrößerung des Halbmessers eines durch Pole begrenzten Konvergenzkreises von Potenzreihen, ZAMM 39 (1959) 12.

Eingereicht am 31. 4. 1981

Anschrift: GYULA DÉRI, Budapest, József A. utca 18, H-1212, Ungarn

Z AMM 62, 409 – 411 (198 2

Z. F. BACZYŃSKI

On Wave Propagation in an Elastic Layer with Kinematic Constraints

1. Introduction

In this paper modelling of wave propagation in an elastic layer is discussed using a three-dimensional continuum theory with internal constraints [1], [2]. The governing equations for a three-dimensional layer are reduced to a system of two-dimensional equations of motion for the material surfaces bounding the layer, and plane harmonic waves are considered in some cases.

2. Basic equations of the model

Model of elastodynamics for a single isotropic homogeneous layer proposed in [2] is characterized by the basic system of field equations which consists of:

The *equations of motion*:

$$N^{KL}_{(a),K} - h^{-1}M^{KL}_{(a),K} + h^{-1}M^{KL}_{(a-1),K} + h^{-1}N^{3L}_{(a)} - h^{-1}N^{3L}_{(a-1)} +$$
$$+ f^{L}_{(a)} = i^{L}_{(a)} , \quad N^{K3}_{(a),K} - h^{-1}M^{K3}_{(a),K} + h^{-1}M^{K3}_{(a-1),K} -$$
$$- h^{-1}N^{33}_{(a)} + h^{-1}N^{33}_{(a-1)} + f^{3}_{(a)} = i^{3}_{(a)} \qquad (1)$$

where

$$M^{ij}_{(a)} = \int_0^h T^{ij}\xi d\xi , \qquad N^{ij}_{(a)} = \int_0^h T^{ij} d\xi , \qquad 0 \leqq \xi \leqq h ,$$
$$a = 1, 2 ; \qquad K, L = 1, 2 ; \qquad i, j = 1, 2, 3 .$$

The *geometrical relations*:

$$e_{(a)MN} = u_{(a)(M,N)} , \qquad e_{(a)M3} = (2h)^{-1} (u_{(a+1)M} - u_{(a)M}) +$$
$$+ 2^{-1}u_{(a)3,M} ,$$
$$e'_{(a)MN} = h^{-1}(u_{(a+1)(M,N)} - u_{(a)(M,N)}) , \qquad (2)$$
$$e'_{(a)M3} = (2h)^{-1} (u_{(a+1)3,M} - u_{(a)3,M}) ,$$
$$e_{(a)33} = h^{-1}(u_{(a+1)3} - u_{(a)3}) .$$

The *constitutive relations*:

$$N^{ij}_{(a)} = I^{ijkl}_{(a)0}e_{(a)kl} + I^{ijkl}_{(a)1}e'_{(a)kl} + I^{ijkl}_{(a-1)0}e_{(a-1)kl} + I^{ijkl}_{(a-1)1}e'_{(a-1)kl} ,$$
$$M^{ij}_{(a)} = I^{ijkl}_{(a)1}e_{(a)kl} + I^{ijkl}_{(a)2}e'_{(a)kl} + I^{ijkl}_{(a-1)1}e_{(a-1)kl} +$$
$$+ I^{ijkl}_{(a-1)2}e'_{(a-1)kl} , \qquad (3)$$

where

$$I^{ijkl}_{(a)n} = \int_0^h C^{ijkl}\xi^n d\xi , \qquad (n = 0, 1, 2) .$$

The *kinetic boundary conditions*:

$$(N^{KL}_{(a)} - h^{-1}M^{KL}_{(a)} + h^{-1}M^{KL}_{(a-1)})\, n_{(a)K} = p^{L}_{(a)} , \qquad (4)$$
$$(N^{K3}_{(a)} - h^{-1}M^{K3}_{(a)} + h^{-1}M^{K3}_{(a-1)})\, n_{(a)K} = p^{3}_{(a)} , \quad \text{for } \overline{X} \in \partial\Pi ;$$

and the *initial conditions*:

$$u_{(a)}(\cdot,0) = u_{(a)0}(\overline{X}) , \qquad \dot{u}_{(a)}(\cdot,0) = u_{(a)1}(\overline{X}) , \quad \text{for } \overline{X} \in \Pi . \quad (5)$$

Here, $M = M(\overline{X}, t)$, $N = N(\overline{X}, t)$, $f = f(\overline{X}, t)$, $i = i(\overline{X}, t)$, $p = p(\overline{X}, t)$, $u = u(\overline{X}, t)$ $e = e(\overline{X}, t)$, $e' = e'(\overline{X}, t)$ denote in this order, the generalized internal and external forces, the inertia forces and boundary loads, displacements and deformations; h denotes the thickness of the layer.

3. Decomposition of displacement vector fields

A starting-point of analysis of the wave propagation problem in the elastic layer are the displacement equations of motion for two-dimensional vector fields $u_{(1)} = u_{(1)}(\overline{X}, t)$, $u_{(2)} = u_{(2)}(\overline{X}, t)$ which describe the motion of material surfaces $\pi_{(1)}$, $\pi_{(2)}$ bounding the layer. The displacement equations of motion take the following form:

$$(\lambda + \mu)\,(u_{(1)K} + u_{(2)K})_{,KL} + \mu(u_{(1)L} + u_{(2)L})_{,KK} +$$
$$+ 2\lambda h^{-1}(u_{(2)3} - u_{(1)3})_{,L} + 2h^{-1}(f_{(1)L} + f_{(2)L})$$
$$= \varrho(\ddot{u}_{(1)L} + \ddot{u}_{(2)L}) ,$$
$$(\lambda + \mu)\,(u_{(1)K} - u_{(2)K})_{,KL} + \mu(u_{(1)L} - u_{(2)L})_{,KK} +$$
$$+ 6\mu h^{-1}(u_{(1)3} + u_{(2)3})_{,L} + 12\mu h^{-2}\,(u_{(2)L} - u_{(1)L}) +$$
$$+ 6h^{-1}(f_{(1)L} - f_{(2)L}) = \varrho(\ddot{u}_{(1)L} - \ddot{u}_{(2)L}) ,$$
$$\mu(u_{(1)3} + u_{(2)3})_{,KK} + 2\mu h^{-1}(u_{(2)K} - u_{(1)K})_{,K} +$$
$$+ 2h^{-1}(f_{(1)3} + f_{(2)3}) = \varrho(\ddot{u}_{(1)3} + \ddot{u}_{(2)3}) ,$$
$$\mu(u_{(1)3} - u_{(2)3})_{,KK} + 6\lambda h^{-1}(u_{(2)K} + u_{(1)K})_{,K} +$$
$$+ 12h^{-2}(\lambda + 2\mu)\,(u_{(2)3} - u_{(1)3}) + 6h^{-1}(f_{(1)3} - f_{(2)3})$$
$$= \varrho(\ddot{u}_{(1)3} - \ddot{u}_{(2)3}) \qquad (6)$$

where λ, μ are the LAMÉ constants.

We introduce the auxiliary vector fields: $v = (v_L, v_3)$, $w = (w_L, w_3)$ characterizing the general symmetric and antisymmetric processes in such a way that:

$$v_L = 2^{-1}(u_{(1)L} + u_{(2)L}) , \qquad v_3 = h^{-1}(u_{(1)3} - u_{(2)3}) ,$$
$$w_L = h^{-1}(u_{(1)L} - u_{(2)L}) , \qquad w_3 = 2^{-1}(u_{(1)3} + u_{(2)3}) . \qquad (7)$$

To these displacement fields there correspond the symmetric and antisymmetric external forces: $g = (g_L, g_3)$, $q = (q_L, q_3)$, so that,

$$g_L = 2^{-1}(f_{(1)L} + f_{(2)L}) , \qquad g_3 = h^{-1}(f_{(1)3} - f_{(2)3}) ,$$
$$q_L = h^{-1}(f_{(1)L} - f_{(2)L}) , \qquad q_3 = 2^{-1}(f_{(1)3} + f_{(2)3}) . \qquad (8)$$

If we use the definitions, then the displacement equations (6) can be decomposed into two independent systems of equations characterizing the generalized symmetr icprocess (compressional), and the generalized antisymmetric process (flexural).

The first system of equations has the form:

$$\mu v_{L,KK} + (\lambda + \mu) v_{K,KL} - \lambda v_{3,L} + 2h^{-1} g_L = \varrho \ddot{v}_L ;$$

$$(\boldsymbol{w} = \boldsymbol{0}, \boldsymbol{q} = \boldsymbol{0})$$

$$\mu v_{3,KK} + 12\lambda h^{-2} v_{K,K} - 12(\lambda + 2\mu) h^{-2} v_3 + 6h^{-1} g_3 = \varrho \ddot{v}_3 \qquad (9)$$

where $v_L = v_L(\overline{X}, t)$, and $v_3 = v_3(\overline{X}, t)$ are two-dimensional symmetric plane and antiplane motions, while $g_L = g_L(\overline{X}, t)$, and $g_3 = g_3(\overline{X}, t)$ are plane and antiplane external forces.

The second system of equations takes the form:

$$\mu w_{L,KK} + (\lambda + \mu) w_{K,KL} + 12\mu h^{-2} w_{3,L} - 12\mu h^{-2} w_L + 6h^{-1} q_L = \varrho \ddot{w}_L ; \quad (\boldsymbol{v} = \boldsymbol{0}, \boldsymbol{g} = \boldsymbol{0})$$

$$\mu w_{3,KK} - \mu w_{K,K} + 2h^{-1} q_3 = \varrho \ddot{w}_3 , \qquad (10)$$

where $w_L = w_L(\overline{X}, t)$, and $w_3 = w_3(\overline{X}, t)$ denote two-dimensional antisymmetric plane and antiplane motions, while $q_L = q_L(\overline{X}, t)$, and $q_3 = q_3(\overline{X}, t)$ are plane and antiplane external forces.

4. Propagation of plane harmonic waves

We consider the propagation of plane harmonic waves in an elastic layer, free from body forces, that is bounded by two parallel planes $X^3 = 0$, $X^3 = h$, and extends to infinity in the $\pm (X^1, X^2)$-directions. We will discuss the following two-dimensional processes: generalized plane symmetric process, and generalized plane antisymmetric process.

Generalized plane symmetric process ($\boldsymbol{g} = \boldsymbol{0}$)

The problem of generalized plane symmetric motions requires the solution of displacement equations in the form:

$$\mu \overset{*}{v}_{L,KK} + (\lambda + \mu) \overset{*}{v}_{K,KL} - \lambda h^{-1} \overset{*}{v}_{3,L} = \varrho \ddot{\overset{*}{v}}_L , \qquad \overset{*}{v}_L \equiv v_L , \; (11)$$

$$\mu \overset{*}{v}_{3,KK} + 12\lambda h^{-1} \overset{*}{v}_{K,K} - 12(\lambda + 2\mu) h^{-2} \overset{*}{v}_3 = \varrho \ddot{\overset{*}{v}}_3 , \qquad \overset{*}{v}_3 \equiv h v_3 .$$

We seek the solution of equation (11) in the form:

$$\overset{*}{\boldsymbol{v}} = \boldsymbol{a} \exp i(\boldsymbol{k} \cdot \overline{\boldsymbol{X}} - \omega t) , \qquad (12)$$

where $\boldsymbol{a}$ and $\boldsymbol{k}$ are constant vectors, and ω is a scalar constant. If $\boldsymbol{n} = \boldsymbol{k}/k$ where $k = |\boldsymbol{k}|$, then the solution (12) represents a plane harmonic wave with amplitude $\boldsymbol{a}$ propagating in the direction $\boldsymbol{n}$ and with constant velocity $c = \omega/k$ where ω is the circular frequency and k is the wave number.

Substituting (12) into (11) we obtain the homogeneous equations for the amplitude vector:

$$\mu k^2 a_L + (\lambda + \mu) k^2 n_K n_L a_K + \lambda i k h^{-1} n_L a_3 - \varrho k^2 c^2 a_L = 0 ,$$

$$\mu k^2 a_3 - 12\lambda i k h^{-1} n_K a_K + 12(\lambda + 2\mu) h^{-2} a_3 - \varrho k^2 c^2 a_3 = 0 . \quad (13)$$

Now, we discuss the wave propagation in case of longitudinal waves, and shear waves.

(i) If we select (X^1, X^2)-plane so that the propagation direction $\boldsymbol{n}$ is in this plane, then the waves are independent of X^3. For longitudinal waves (P-waves), the displacement vector is perpendicular to the plane wave front and is always in the (X^1, X^2)-plane. We have: $\boldsymbol{a} \| \boldsymbol{n}$, i.e. $\boldsymbol{a} \times \boldsymbol{n} = \boldsymbol{0}$, $\boldsymbol{a} = \eta \boldsymbol{n}$, $n_3 = 0$, $a_3 = 0$, and from (13) we obtain the following system of equations for the amplitude vector:

$$\eta[\mu \varrho^{-1} n_L + (\lambda + \mu) \varrho^{-1} n_K n_L n_K - c^2 n_L] = 0 ,$$

$$12\lambda i (kh)^{-1} n_K n_K \eta = 0 . \qquad (14)$$

This system of equations has only trivial solutions because $n_K n_K = 1$, and $0 < h < \infty$, therefore the propagation of plane longitudinal waves on bounding surfaces of the elastic layer with finite thickness is not possible. In the case, when the product $kh \to \infty$, the plane longitudinal waves can propagate with classical velocity c_1, and plane harmonic motions have the form:

$$\text{P-waves: } \overset{*}{v}_L(\overline{X}, t) = \overset{*}{a}_L \exp i(n_K X_K - c_1 t) ,$$

$$\varepsilon_{KL} n_L \overset{*}{a}_K = 0 ; \qquad \overset{*}{v}_3(\overline{X}, t) = 0 . \qquad (15)$$

(ii) For the shear waves (S-waves), the displacement vector is in the plane wave front. The displacement vector of the S-waves can be decomposed into two components: one in the (X^1, X^2)-plane as a horizontally polarized shear wave (SH-waves), and the other perpendicular to it as a vertically polarized shear wave (SV-waves). For the shear waves we have: $\boldsymbol{a} | \boldsymbol{n}$, i.e. $\boldsymbol{a} \cdot \boldsymbol{n} = 0$, $n_3 = 0$, $a_K n_K = 0$, and from (13) we obtain the following system of equations for the amplitude vector:

$$(\mu \varrho^{-1} - c^2) a_L + \lambda i (\varrho k h)^{-1} n_L a_3 = 0 ,$$

$$[\mu \varrho^{-1} - c^2 + 12(\lambda + 2\mu) (kh)^{-2} \varrho^{-1}] a_3 = 0 . \qquad (16)$$

This system of equations can be satisfied if and only if $a_3 = 0$, then we have one wave velocity $c_2 = (\mu/\varrho)^{1/2}$ which is classical velocity of propagation for plane harmonic horizontally polarized waves:

$$\text{SH-waves: } \overset{*}{v}_L(\overline{X}, t) = \overset{*}{a}_L \exp i(n_K X_K - c_2 t) ,$$

$$\overset{*}{a}_L n_L = 0 ; \qquad \overset{*}{v}_3(\overline{X}, t) = 0 . \qquad (17)$$

Generalizedplane antisymmetric process ($\boldsymbol{q} = \boldsymbol{0}$)

The problem of generalized plane antisymmetric motions requires the solution of the displacement equations:

$$\mu w_{L,KK} + (\lambda + \mu) \overset{*}{w}_{K,KL} + 12\mu h^{-3} \overset{*}{w}_{3,L} - 12\mu h^{-2} \overset{*}{w}_L = \varrho \ddot{\overset{*}{w}}_L ,$$

$$\overset{*}{w}_L \equiv w_L , \quad \mu \overset{*}{w}_{3,KK} - \mu h \overset{*}{w}_{K,K} = \varrho \ddot{\overset{*}{w}}_3 , \qquad \overset{*}{w}_3 \equiv h w_3 . \qquad (18)$$

We seek a solution of equation (17) in the form similar to the plane symmetric process:

$$\overset{*}{\boldsymbol{w}} = \boldsymbol{a} \exp i(\boldsymbol{k} \cdot \overline{\boldsymbol{X}} - \omega t) , \qquad (19)$$

and from (18) we obtain the following system of homogeneous equations for the amplitude vector:

$$\mu k^2 n_K n_K a_L + (\lambda + \mu) k^2 n_K n_L a_K + 12\mu i k h^{-3} n_L a_3 - 12\mu h^{-2} a_L - \varrho k^2 c^2 a_L = 0 ,$$

$$\mu k^2 n_K n_K a_3 - \mu i k h n_K a_K - \varrho k^2 c^2 a_3 = 0 . \qquad (20)$$

Let us discuss the wave propagation in case of longitudinal waves and shear waves.

(i) For longitudinal waves (P-waves) we have: $\boldsymbol{a} \| \boldsymbol{n}$, $\boldsymbol{a} \times \boldsymbol{n} = \boldsymbol{0}$, $\boldsymbol{a} = \eta \boldsymbol{n}$, $n_3 = 0$, $a_3 = 0$, and from (20) we obtain:

$$\eta[(\lambda + 2\mu) \varrho^{-1} + 12\mu (kh)^{-2} \varrho^{-1} - c^2] = 0 ,$$

$$\eta \mu i (kh)^{-1} n_K n_K = 0 . \qquad (21)$$

This system of equations has only zero solutions, therefore the propagation of plane longitudinal waves on bounding surfaces of the elastic layer with finite thickness is not possible. In the case, when the $kh \to \infty$, the plane longitudinal waves can propagate with classical velocity c_1, and plane harmonic motions have the form:

$$\text{P-waves: } \overset{*}{w}_L(\overline{X}, t) = \overset{*}{a}_L \exp i(n_K X_K - c_1 t) ,$$

$$\varepsilon_{KL} n_L \overset{*}{a}_K = 0 ; \qquad \overset{*}{w}_3(\overline{X}, t) = 0 . \qquad (22)$$

(ii) For shear waves (S-waves) we have: $\boldsymbol{a} \perp \boldsymbol{n}$, $\boldsymbol{a} \cdot \boldsymbol{n} = 0$, $n_3 = 0$, $a_K n_K = 0$, and from (20) we obtain:

$$[\mu \varrho^{-1} - c^2 + 12\mu (kh)^{-2} \varrho^{-1}] a_L + 12\mu i (\varrho k)^{-1} h^{-3} n_L n_3 = 0 ,$$

$$(\mu \varrho^{-1} - c^2) a_3 = 0 . \qquad (23)$$

This system of equations can be satisfied if and only if $a_3 = 0$, then we have one wave velocity: $c = \overset{*}{c}_2 = c_2[1 + 12(kh)^{-2}]^{1/2}$, and plane harmonic horizontally polarized motions have the form:

$$\text{SH-waves: } \overset{*}{w}_L(\overline{X}, t) = \overset{*}{a}_L \exp i(n_K X_K - \overset{*}{c}_2 t) ,$$

$$\overset{*}{a}_L n_L = 0 ; \qquad \overset{*}{w}_3(\overline{X}, t) = 0 . \qquad (24)$$

In the case, when the $kh \to \infty$, then the plane harmonic SH-waves can propagate with classical velocity c_2.

To obtain motion of a particle lying in interior of the layer we use the formulae:

$$\boldsymbol{u}(X, t) = \boldsymbol{u}_{(1)}(\bar{X}, t) + \xi h^{-1}[\boldsymbol{u}_{(2)}(\bar{X}, t) - \boldsymbol{u}_{(1)}(\bar{X}, t)], \quad 0 \leqq \xi \leqq h,$$

$$X = (X^1, X^2, X^3) \in \Pi \times (0, h), \qquad \bar{X} = (X^1, X^2) \in \Pi,$$

$$t \in T. \tag{25}$$

The propagation conditions that we have found for the two-dimensional motions $\boldsymbol{u}_{(1)}$, and $\boldsymbol{u}_{(2)}$ imply suitable restrictions on the three-dimensional fields $\boldsymbol{u} = \boldsymbol{u}(X, t)$.

References

1 Woźniak, C., On the relation between discrete and continuum mechanics of certain material system, Arch. Mech. Stos., 28, 3, pp. 571–581 (1976).

2 Baczyński, Z. F., Mathematical models of elastodynamics of the layered bodies (to be published).

Eingereicht am 2. 7. 1981 (Vortrag GAMM-Tagung Würzburg 1981)

Anschrift: Dr.-Ing. Z. F. Baczyński, Polish Academy of Sciences, Institute of Fundamental Technological Research, Świętokrzyska 21, 00-049 Warszawa, Poland.

ZAMM 62, 411–415 (1982)

N. K. Gupta / B. P. Ambasht

On Non-Uniform Deformation in Tension: An Experimental Study

Introduction

The material presented in this paper is the result of a series of experiments performed to obtain and analyse the large deformation data with a view to study the process of transition from uniform to nonuniform deformation in tension. The study has generated some interesting results which are relevant to three basic aspects of the phenomenon viz; (a) the stress and load conditions at the onset of nonuniform deformation and its further localisation; (b) the manner in which uniform deformation transforms into nonuniform deformation; and (c) the effect of specimen size on the response of material both before and after the onset of necking. The observations reveal that the result of some recent studies, that the ratio of true stress at the point of necking and true stress at maximum load depends on specimen length to diameter ratio, for example [1], is not fully substantiated.

Experimental

Axisymmetric round specimens of seven different sizes of aluminium (aluminium 99.5%, copper 0.04%) and three different sizes (table 1) of copper (copper 99.9%, bismuth 0.01%, iron 0.05%) were machined from 50 mm diameter rolled rods, obtained from single batch of their respective production. These specimens were annealed for two hours at 400 °C for aluminium and 450 °C for copper and were furnace cooled before they were tested in tension.

The initiation of nonuniform deformation in the test was identified by taking simultaneous measurements of diameters at two sections of the specimen; viz, the neck section and a section away from it. To be able to make it possible that the mid section of specimen is a favoured section for the neck to appear there, it was reduced in diameter by 0.25%. A set of preliminary experiments was conducted on specimens whose diameters were systematically reduced by small amounts at the middle of their lengths. The test results and the frequency of appearance of neck at the reduced section were compaired with those for a normal specimen. A reduction of 0.25% was found to give desired results because four out of five specimens necked at the middle section, while there was no effect on load deformation or the stress strain behaviour of the material.

Three specimens of each geometry of aluminium and two each of copper were tested up to each of five different stages of necking including fracture in a 50 T Avery universal testing machine at a crosshead speed of about 0.01 mm/s. The simultaneous measurements taken at small intervals were those of (a) increments in applied load; (b) corresponding increase in specimen length; and (c) diameter readings at two sections of the specimen, viz, at the neck section and a section away from it. For measuring the diameters, special diameter indicators were developed; and the manner in which they were mounted on the specimen is shown in Fig. 1. It may be seen in this figure that the knife edges (1 and 2) are respectively fixed to frame (3) and dial gauge (4). Checked by spirit level (7), the stem of the dial gauge is kept horizontal during the test by small rotation of levelling screw (6). Adjusting nut (9) is used to vary the stiffness of spring (8) such that the extensometer is always held to the specimen with a light pressure. The end of levelling screw rests against a phosphor bronze guide (10) clipped to the specimen.

Results and Discussions

(a) Initiation of Nonuniform Deformation

The load versus diameter readings, at specimen middle section and a section away from it, were superposed on one another for each specimen of both materials. Plots for typical specimens of aluminium and copper are shown in Figs. 2 and 3 respectively, wherein same numbers correspond to diameter readings at two sections for same value of the load. It is seen that with load rising appreciably, these diameter readings initially coincide with each other, indicating thereby that the deformation is uniform. As the load approaches its maximum value, the readings at the neck section begin to show more deformation there than at section away from it. Nonuniform deformation, therefore, seems to have initiated at this point, which appears to be near reading 6 or 9 of Fig. 2 or 3. This process seems to continue until section away from the neck ceases to register any further deformation (around reading 9 or 14 of Fig. 2 or 3) and a localised neck begins to form, while the applied load begins to show a fall.

(b) Load Deformation Curves

The load deformation plots for typical specimens of aluminium and copper are shown in Figs. 4 and 5 respectively, wherein points B and C correspond to the initiation of nonuniform deformation and beginning of localised necking respectively. Point D corresponds to fracture. It is interesting to note that the load variations from B to C are rather small. For the sake of further discussions, however, we approximate these graphs to be composed of three regions, see also [3], viz; (1) region A to B, during which load rise is appreciable; (2) region B to C, during which the load rise is rather negligible; and (3) region C to D, during which load is falling. The extension of specimen during these periods is designated by δ_1, δ_2 and δ_3 (see Figs. 4 and 5); the total extension of the specimen at fracture, therefore, is

$$\delta = \delta_1 + \delta_2 + \delta_3 .$$

Values of δ_1, δ_2 and δ_3 were measured for each specimen in order to study their dependence on specimen size. It was found that while δ_1 shows a linear dependence on original length (L_0) of the specimen, as shown in Figs. 6 and 7, it appears to be independent of the original diameter (d_0), as shown in Figs. 8 and 9. It is interesting to note that δ_2 appears to have dependence on both L_0 and d_0 of the specimen, see Figs. 8 and 9, while δ_3 shows dependence on d_0 only, see Figs. 10 and 11, being rather independent of L_0. These observations seem to justify the assumption made by Onat and Prager [2] regarding the length, over which the neck forms, to be a function of d_0.

(c) Stress Strain Curves

Stress, P/A, and strain, $\ln(A_0/A)$, values were plotted for all specimens, as shown in Figs. 12 and 13, for typical series of specimens of aluminium and copper respectively. Here, points a, b, c and d correspond to points A, B, C and D of Fig. 4 or 5.

A remarkable feature revealed by these plots is that stress value corresponding to point a, (σ_{i1}), at which nonuniform deformation begins, or point b, (σ_{i2}), at which localisation of necking begins, is rather independent of the original length or diameter of specimen. Both these values appear not to vary within the range of present experiments and they come out to be, for aluminium

$\sigma_{i1} = 101.0\ \mathrm{MN\ m^{-2}}$, $\sigma_{i2} = 116.7\ \mathrm{MN\ m^{-2}}$ and for copper

$\sigma_{i1} = 263.9\ \mathrm{MN\ m^{-2}}$, $\sigma_{i2} = 298.2\ \mathrm{MN\ m^{-2}}$.

An analysis of nonuniform deformation, details of which are being presented elsewhere, was developed [4] by approximating the neck profile by a relationship of the form:

$$r(X) = F - G \exp(-HX/r_0) \quad \text{for} \quad X \geqq 0 ,$$

$$r(X) = r(-X) ,$$

X, being the material coordinate and F, G and H the functions of strain.

The stress strain relationship used in the analysis is of the form:

$$\varepsilon = \frac{\sigma}{E} + \frac{E - \sigma_{i1}}{n_1 E} \left(\frac{\sigma}{\sigma_{i1}}\right)^{n_1}, \quad \text{for region ab (Fig. 12)}$$

$$\varepsilon = \ln \frac{\sigma}{\sigma_{i1}} + \frac{E + (n_1 - 1)\,\sigma_{i1}}{n_1 E}, \quad \text{for region bc,}$$

$$\varepsilon = \frac{\sigma}{E} + \frac{E - \sigma_{i2}}{n_2 E} \left(\frac{\sigma}{\sigma_{i2}}\right)^{n_2}, \quad \text{for region cd.}$$

Solving for extension, a close form solution for δ_2 shows that

$$\delta_2 = f(L_0/d_0)\, d_0 ,$$

and for $L_0/d_0 = \text{const.}\ \delta_2 = \alpha d_0$.

Numerical integration for the solution of δ_3, however, showed that it depended on d_0, and the influence of L_0 on it comes out to be negligible. Fig. 14 shows the nature of variation of δ_2 with d_0 for different values of L_0/d_0, in case of copper specimens. The values of n_1 and n_2, taken for these computations, were 4.39 and 2.86 respectively.

Conclusions

The results of present study indicate that the initiation of nonuniform deformation appears to occur earlier than the point at which load having reached its maximum begins to fall. There exists a transition region (B – C of Fig. 4 or 5 and b – c of Fig. 12 or 13,) during which period the uniform deformation transforms gradually into localised necking.

The data obtained show that the specimen size influences the load extension plot, while the stress strain results are not effected by it.

Acknowledgement

The paper was written while Dr. GUPTA held ALEXANDER-VON-HUMBOLDT-Stiftung fellowship at the Institut für Mechanik, Ruhr-Universität Bochum.

Table 1. Specimen Dimensions

SNo	Speci-men	d_0	L_0	SNo	Speci-men	d_0	L_0
1	A_0A	30	200	6	A_3	10	140
2	A_0B	30	150	7	A_4	6	100
3	A_0C	30	100	8	C_1	22	100
4	A_1	22	100	9	C_2	15	140
5	A_2	15	140	10	C_3	10	140

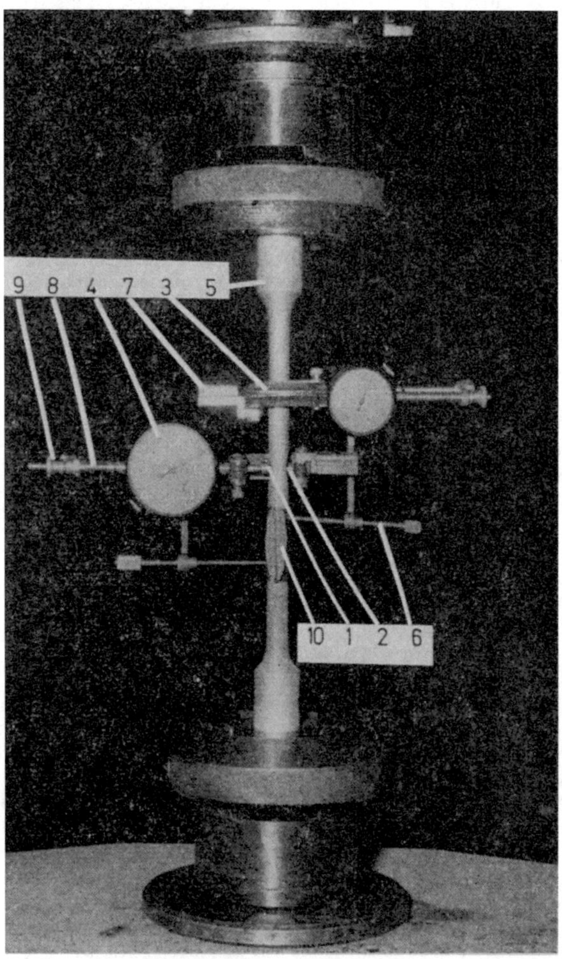

Fig. 1. Testing arrangement

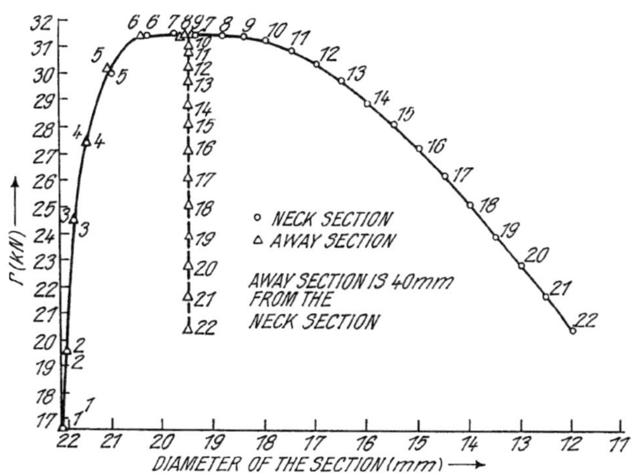

Fig. 2. Load versus diameter plot for a typical specimen of aluminium (A1 – 4)

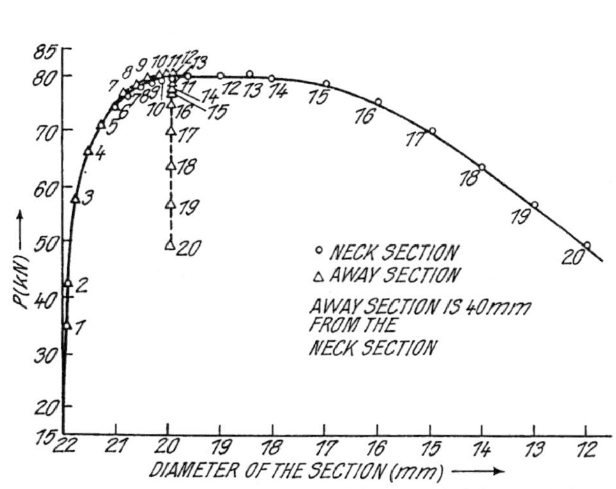

Fig. 3. Load versus diameter plot for a typical specimen of copper (C1 – 1)

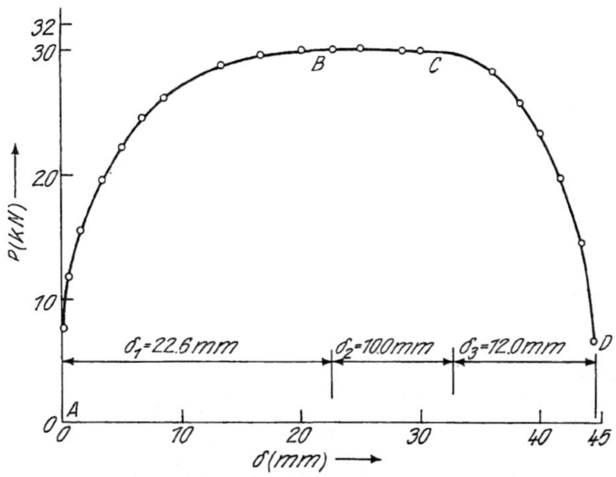

Fig. 4. Load extension plot for an aluminium specimen (Al – 2)

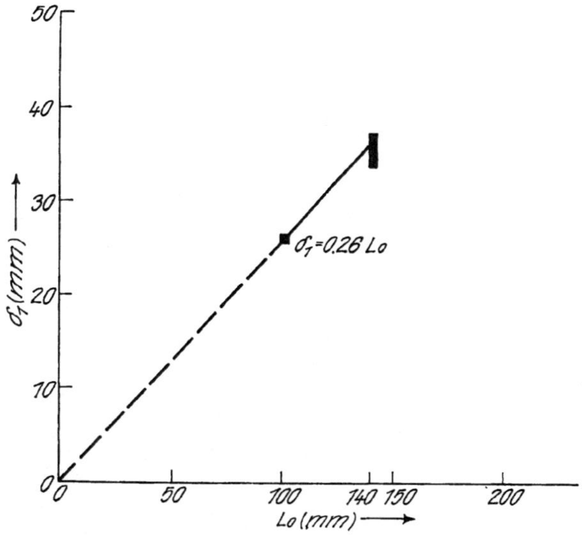

Fig. 7. Variation of δ_1 with L_0 for copper

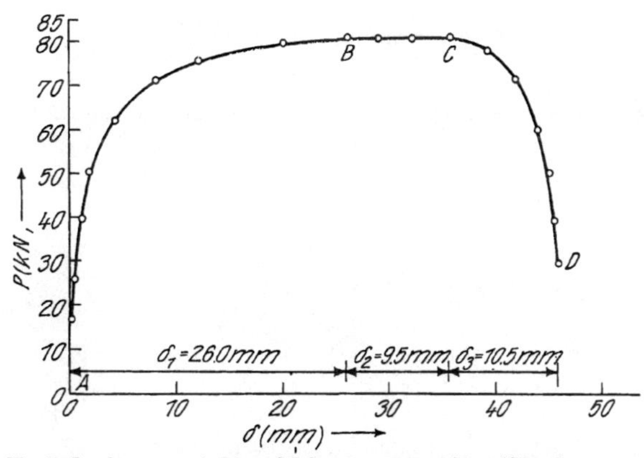

Fig. 5. Load versus extension plot for a copper specimen (C1 – 1)

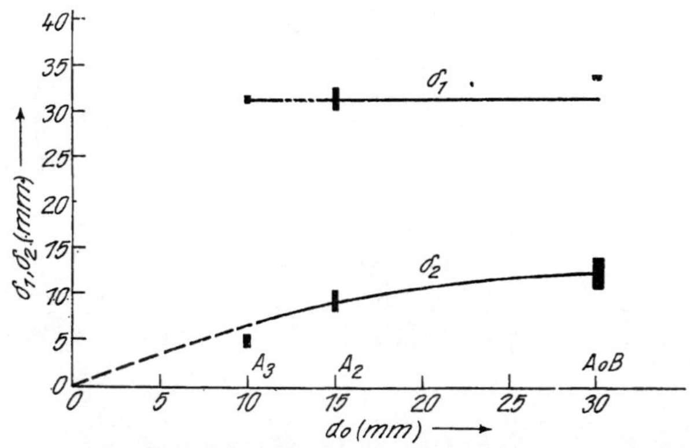

Fig. 8. Variation of δ_1 and δ_2 with d_0 for aluminium specimens of L_0 = 140 mm

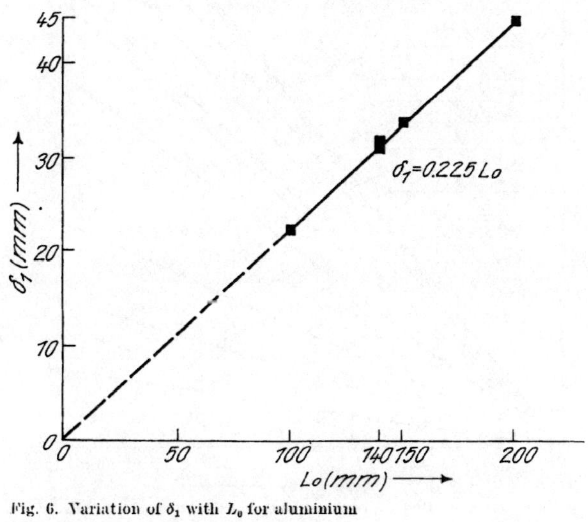

Fig. 6. Variation of δ_1 with L_0 for aluminium

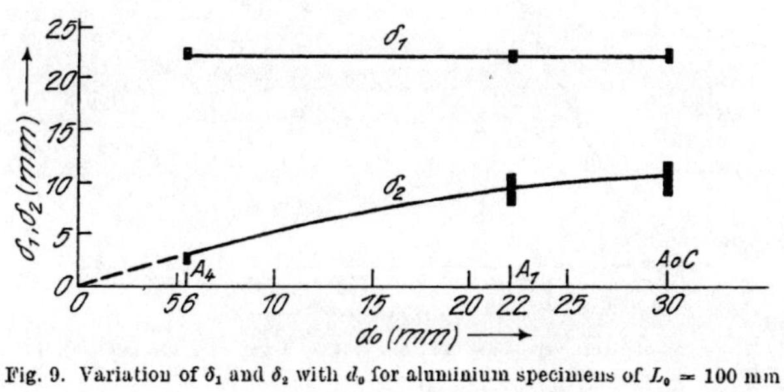

Fig. 9. Variation of δ_1 and δ_2 with d_0 for aluminium specimens of L_0 = 100 mm

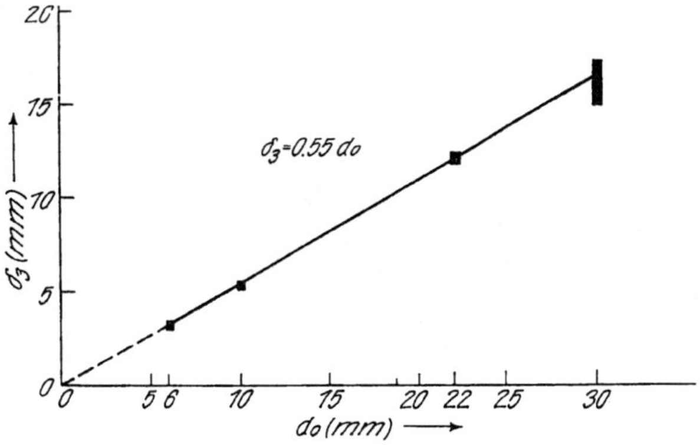

Fig. 10. Variation of δ_3 with d_0 for aluminium specimens

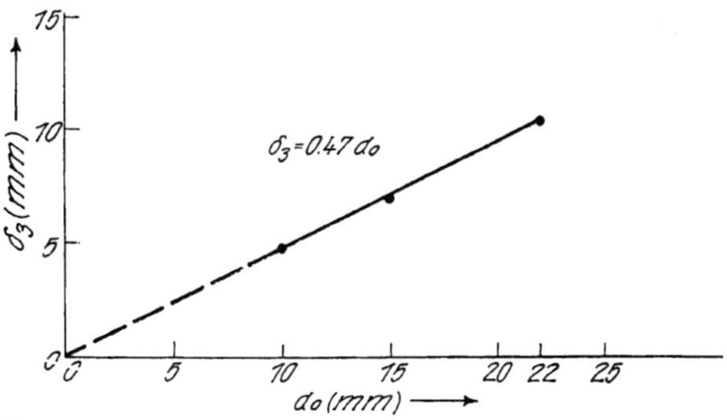

Fig. 11. Variation of δ_3 with d_0 for copper specimens

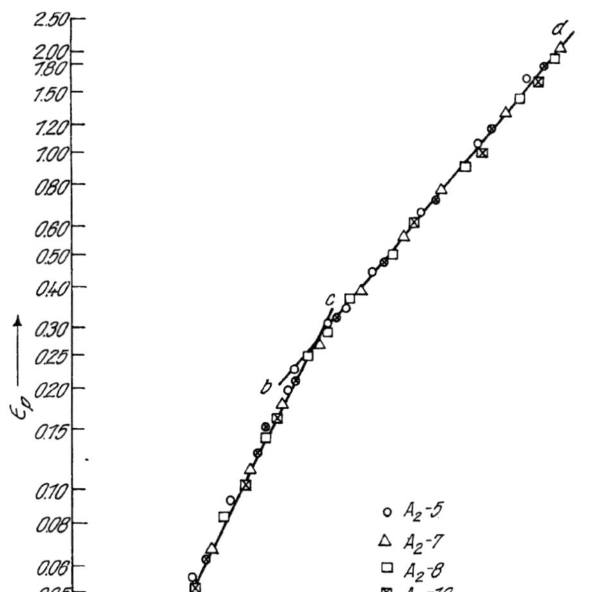

Fig. 12. A log of stress versus log of strain plot for aluminium specimens of A2 series

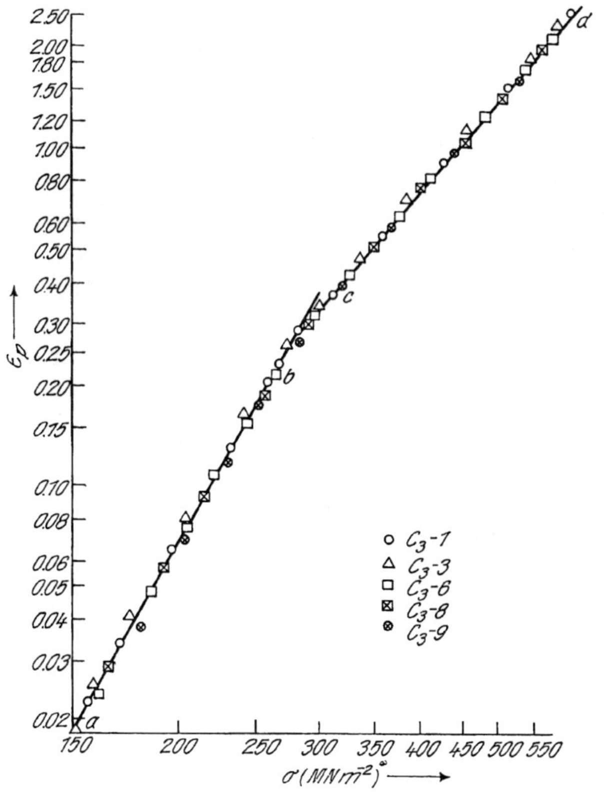

Fig. 13. A log of stress versus log of strain plot for copper specimens of C3 series

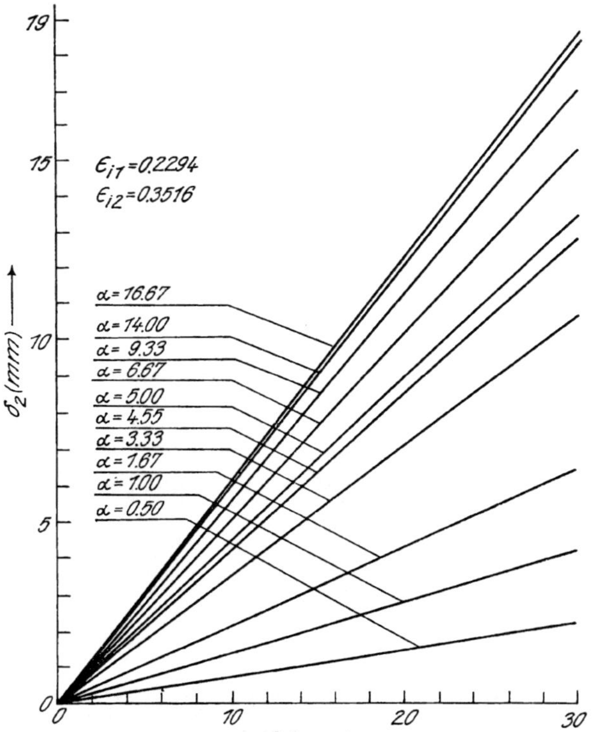

Fig. 14. Variation of δ_2 with d_0, for constant α ($= L_0/d_0$), for specimens of copper

References

1 HUTCHINSON, J. W.; MILES, J. P., Bifurcation analysis of the onset of necking in an elastic/plastic cylinder under uniaxial tension, J. Mech. Phys. Solids, vol 22, pp 61–71 (1974).
2 ONAT, E. T.; PRAGER, W., The necking of a tension specimen in plane plastic flow, Report, Watertown Arsenal No DA-2598/2 (1953).
3 GILLIS, P. P., JONES, S. E., Tensile deformation of a round bar, J. Engg. Maths., vol 12, No 4 (1978).
4 AMBASHT, B. P., Large deformation including necking under uniaxial tension, Ph. D. Thesis, I. I. T., Delhi (1979).

Eingereicht am 25. 5. 1981 (Vortrag GAMM-Tagung Würzburg 1981)

Address: Dr. N. K. GUPTA, Department of Applied Mechanics, Indian Institute of Technology, New Delhi, India / Dr. B. P. AMBASHT, Department of Mechanical Engineering, Bihar Institute of Technology, Sindri, India

ZAMM 62, 415–417 (1982)

H. PÉREZ-DE-TEJADA / M. S. UBEROI

Magnetohydrodynamic Boundary Layer

Introduction

Viscous boundary layer theory is, in general, characterized by the concept that at high values of the REYNOLDS number, the motion of a streaming fluid is controlled by viscous forces only within a thin region adjacent to the walls that are in contact with the fluid. Under appropriate conditions the bulk of the flow remains unaffected by this action and, to a large extent, is unaware of the existence of the boundary layer. In fact, the small velocity perturbation forced outside of the viscous flow region is known to exhibit a characteristic decrease with the downstream distance from the leading edge.

The application of this notion to MHD boundary layer theory must bear inherently the same geometric and dynamic constraints if the boundary layer concept itself is to remain valid. Within the context of this basic consideration it is, therefore, necessary to realize that additional restrictions must be imposed in this case in order to exclude the electrodynamic influence of the viscous flow region from the rest of the flow. Most important is the recognition that the integrated electric current across the boundary layer must be identically zero. In fact, the violation of this restriction would result in the generation of a perturbation magnetic field, outside of the boundary layer, whose intensity and geometry is determined primarily by the dimensions of the wall and the boundary conditions at the end plates.

The validity of the zero integrated electric current as a necessary condition in MHD boundary layer theory has been assumed implicitly, in the past, in studies of the motion of a conducting flow streaming over both sides of a thin-semi-infinite flat plate (GREENSPAN and CARRIER, 1959). Since in this case (shown in figure 1a) the flow produces symmetric profiles for both the velocity and the magnetic fields, the integrated electric current across both boundary layers is identically zero and hence the magnetic field profile is confined to a narrow region. On the other hand, since in the boundary layer formed over one side of a semi-infinite flat non-conducting wall this condition must be given explicitly, the induced current system associated with this geometry should reverse direction with distance away from the wall as indicated in figure 1b. In the case of a current sheet on a perfectly conducting plate or of an electric current imposed externally on the plate, the total current across the layer(s) plus the current in the plate must vanish in order for the boundary layer(s) to exist. In the present study we examine the boundary layer formed over one side of a semi-infinite non-conducting wall by using the similarity analysis of GREENSPAN-CARRIER for magnetic-field aligned geometries. The flow configuration will apply to $x \gg 1$ locations so that leading edge effects resulting, for example, from the removal of the flow at $y < 0$ are neglected.

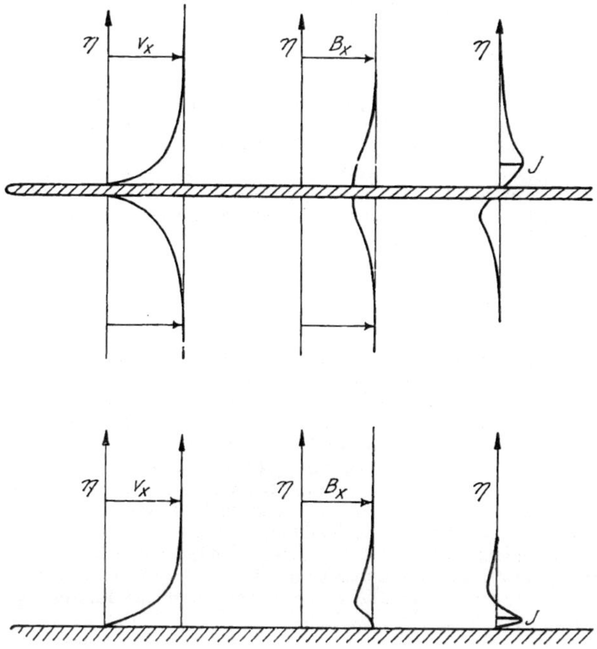

Fig. 1. Schematic diagrams of the velocity, magnetic field and electric current boundary layer profiles of a MHD flow past a thin semi-infinite flat plate (a), and a semi-infinite flat wall (b)

The set of differential equations which governs the steady state motion of an incompressible, magnetic field aligned, electrically conducting viscous flow in semi-infinite geometries is given by:

$$2f''' + ff'' - \beta gg'' = 0\,, \qquad 2g'' + \varepsilon(fg' - f'g) = 0 \tag{1}$$

(see ROBERTS, 1967, and CABANNES, 1970, for the detailed derivation of these equations). f and g denote here functions related to the stream function ψ and the magnetic potential A through:

$$\psi(x, \eta) = -f(\eta) \sqrt{\frac{\hat{x}}{\mathrm{Re}}}\,, \qquad A(x, \eta) = -g(\eta) \sqrt{\frac{\hat{x}}{\mathrm{Re}}} \tag{2}$$

where $\eta = \hat{y} \sqrt{\mathrm{R_e}/\hat{x}}$ represents a suitable similarity variable defined in terms of non-dimensional coordinates $\hat{x}$ and $\hat{y}$, and Re the REYNOLDS number ($\mathrm{Re} = UL/\nu$ with L and U indicating the characteristic dimension and velocity of the flow, and ν the kinematic viscosity coefficient). The parameter β denotes the square of the inverse of the ALFVEN's MACH number and ε the magnetic PRANDTL number given by: $\varepsilon = \mu\sigma\nu$ (μ and σ are the magnetic permeability and the electric conductivity coefficient). As indicated above, the requirement that the integrated electric current across the boundary layer is zero imposes specific geometric restrictions on the electric current and magnetic field profiles. The physical implications of this condition can be better exposed by writing the integrated electric current in terms of similarity variables. We have:

$$\int_0^\infty J \,\mathrm{d}y \simeq \frac{1}{\mu} \int_0^\infty \frac{\partial B_x}{\partial y} \,\mathrm{d}y \sim \frac{1}{\mu} \int_0^\infty g''(\eta) \,\mathrm{d}\eta = \frac{1}{\mu} \left[g'(\infty) - g_0'\right] \tag{3}$$

where $g_0' \equiv g'(0)$. From this expression we see that the restriction of a zero integrated current implies that the horizontal magnetic field component $B_x \sim g'(\eta)$ take identical values at the wall and in the potential flow. In dimensionless form, this is: $g'(0) = g'(\infty) = 1$. These two boundary conditions, together with those for the velocity field, namely $f(0) = f'(0) = 0$ and $f'(\infty) = 1$, define uniquely the solution of equations (1) for different values of the ε and β parameters. An immediate consequence of the condition $g_0' = 1$ is that the value of the magnetic potential at the wall g_0 should no longer be zero (as in the case examined by GREENSPAN and CARRIER), but is determined from the solution of the problem. Thus, non-zero values of g_0 will be necessary to comply with the flow conditions at infinity. Since such values are directly related to that of the transverse component of the magnetic field vector at the wall through:

$$B_y(0) = -\frac{1}{2(\mathrm{Re}\hat{x})^{1/2}} g_0 \tag{4}$$

the similarity solution will require a certain tilting of the magnetic field lines along and beneath the non-conducting wall. This tilting, which is commensurate with that resulting from the displacement of streamlines in the freestream flow, represents a second order effect in the dynamics of the boundary layer flow and its magnitude decreases with increasing distance from the leading edge.

Calculation of the Magnetic Potential at the Wall

The existence of a small transverse component of the magnetic field along the wall is a constraint which is independent of the magnitude of the imposed magnetic field or of the conductivity of the flow. Thus, even in the limiting case where $\beta \to 0$, that is, if the intensity of the applied magnetic field decreases to zero, a non-zero magnetic potential at the wall should be expected. This restriction ensures the necessary tilting of the magnetic field vector to achieve alignment with the (viscously displaced) velocity vector outside the boundary layer. An analytical estimate of the magnetic potential at the wall expected under a weak magnetic field can be elaborated for the particular case in which $\varepsilon = 1$. The convenience of this case can be better exposed by writing equations (1) in the form:

$$\begin{aligned} &8R''' + (1-\varepsilon)\, RS'' + (1+\varepsilon)\, SR'' = 0\,, \\ &8S''' + (1+\varepsilon)\, RS'' + (1-\varepsilon)\, SR'' = 0 \end{aligned} \tag{5}$$

where $S = 2(f + \beta^{1/2} g)$ and $R = 2(f - \beta^{1/2} g)$ (see, for example, CABANNES, 1970). When $\varepsilon = 1$ these equations can be readily integrated to give:

$$\begin{aligned} S''(\eta) &= S''(0)\exp\left(-\frac{1}{4}\int R\,\mathrm{d}\eta\right), \\ R''(\eta) &= R''(0)\exp\left(-\frac{1}{4}\int S\,\mathrm{d}\eta\right). \end{aligned} \tag{6}$$

This integral of the GREENSPAN CARRIER equations is particularly useful to the evaluation of the parameters α and g_0 of the power series:

$$\begin{aligned} f(\eta) &= \frac{\alpha}{2}\eta^2 + \frac{\beta}{4}\frac{\alpha g_0^2}{4!}\eta^4 + \frac{1}{5!}\left(\frac{3}{4}\alpha\beta g_0 - \frac{\alpha^2}{2}\right)\eta^5 + \dots \\ g(\eta) &= g_0 + \eta + \frac{\alpha}{12} g_0 \eta^3 + \frac{\alpha}{48}\eta^4 + \dots \end{aligned} \tag{7}$$

which satisfy equations (1) and the boundary conditions $f(0) = f'(0) = 0$ and $g'(0) = 1$. The adoption of the supplementary boundary conditions $f'(\infty) = g'(\infty) = 1$ can be carried out by integrating from $\eta = 0$ to $\eta = \infty$ the sum and the difference of equations (6). This gives:

$$\begin{aligned} 2 &= \alpha\int_0^\infty \left\{\exp\left(\frac{\beta^{1/2}}{2}\int g\,\mathrm{d}\eta\right) + \exp\left(-\frac{\beta^{1/2}}{2}\int g\,\mathrm{d}\eta\right)\right\} \times \\ &\quad \times \exp\left(-\int\frac{f}{2}\,\mathrm{d}\eta\right)\mathrm{d}\eta\,, \\ 0 &= \int_0^\infty \left\{\exp\left(\frac{\beta^{1/2}}{2}\int g\,\mathrm{d}\eta\right) - \exp\left(-\frac{\beta^{1/2}}{2}\int g\,\mathrm{d}\eta\right)\right\} \times \\ &\quad \times \exp\left(-\int\frac{f}{2}\,\mathrm{d}\eta\right)\mathrm{d}\eta\,. \end{aligned} \tag{8}$$

By employing both power series (7) these expressions can be used to determine the value of the parameters α and g_0 which serve to define, respectively, the skin friction and the magnetic potential at the wall. In order to do this, we must first simplify the integrands of (8) by taking:

$$\exp\left(\pm\frac{\beta^{1/2}}{2}\int g\,\mathrm{d}\eta\right) = 1 \pm \frac{\beta^{1/2}}{2}\int g\,\mathrm{d}\eta + \frac{\beta}{8}\left(\int g\,\mathrm{d}\eta\right)^2 + \dots \tag{9}$$

so that to first order of approximation in β equations (8) reduce to:

$$\begin{aligned} 1 &= \alpha\int_0^\infty \exp\left(-\int\frac{f}{2}\,\mathrm{d}\eta\right)\mathrm{d}\eta, \\ 0 &= \int_0^\infty \left[\int g\,\mathrm{d}\eta\right]\exp\left(-\int\frac{f}{2}\,\mathrm{d}\eta\right)\mathrm{d}\eta\,. \end{aligned} \tag{10}$$

Since to this approximation equation (10a) coincides with that derived directly from equation (1a) when $\beta = 0$, it is permissible to take this limit in the evaluation of the integrands of both equations (10), i.e.:

$$\begin{aligned} \tau &= \int\frac{f}{2}\,\mathrm{d}\eta = \frac{\alpha}{12}\eta^3 - \frac{\alpha^2}{4}\frac{1}{6!}\eta^6 + \frac{11}{8}\frac{\alpha^3}{9!}\eta^9 + \dots, \\ \int g\,\mathrm{d}\eta &= g_0\eta + \frac{1}{2}\eta^2 + \frac{\alpha}{48}g_0\eta^4 + \frac{\alpha}{2}\frac{1}{5!}\eta^5 + \dots\,. \end{aligned} \tag{11}$$

The evaluation of the integrals (10) upon substitution of the power series (11) can be performed by expressing them in terms of the τ variable. Using MEKSYN's procedure (MEKSYN, 1961), we write:

$$\eta = \left(\frac{12}{\alpha}\right)^{1/3}\tau^{1/3}\left\{1 + \frac{\tau}{60} - \frac{\tau^2}{1260} + \dots\right\} \tag{12}$$

so that we now have:

$$\begin{aligned} 1 &= \left(\frac{12\alpha^2}{27}\right)^{1/3}\int_0^\infty\left\{\tau^{-2/3} + \frac{\tau^{1/3}}{15} - \frac{\tau^{4/3}}{180} + \dots\right\}\exp(-\tau)\,\mathrm{d}\tau\,, \\ 0 &= \int_0^\infty\left\{g_0\left(\tau^{-1/3} + \frac{1}{3}\tau^{2/3} + \dots\right) + \right. \\ &\quad \left. + \frac{1}{2}\left(\frac{12}{\alpha}\right)^{1/3}\left(1 + \frac{\tau}{5} + \frac{\tau^2}{100} + \dots\right)\right\}\exp(-\tau)\,\mathrm{d}\tau\,. \end{aligned} \tag{13}$$

These integrals can be evaluated in terms of gamma functions so that we finally have:

$$1 = 2.732\left(\frac{12\alpha^2}{27}\right)^{1/3}, \qquad g_0 = -0.362\left(\frac{12}{\alpha}\right)^{1/3}. \tag{14}$$

The solution of equation (14a) gives the standard $\alpha = 0.33$ value of the non-conducting boundary layer problem which in the present analysis corresponds to the limiting case $\beta = 0$. In this limit the similarity value of the magnetic potential at the wall, given from equation (14b) is $g_0 = -1.18$ so that the transverse component of the magnetic field vector along the wall as given by equation (4) is positive, making the magnetic field lines diverge away from the wall in the direction of the flow. The value of the g_0 parameter derived above can also be used to calculate, from (7b), the profile of the horizontal component of the magnetic field vector given by $B_x \sim g'(\eta)$. Since g_0 is a negative quantity, $g'(\eta)$ should exhibit an initial decrease to the (normalized) minimum value $g'(\eta^*) \simeq 0.81$ at $\eta^* \simeq 2.4$. At this location in the similarity plane, the electric current system reverses direction, forcing the $g'(\eta)$ function to increase again to its freestream value as $\eta \to \infty$. Even though this description can strictly be applied only to the limiting case $\beta = 0$ it should also hold adequately in cases in which β is small. Such is in fact indicated by the results of the numerical integration.

Results of the Numerical Integration

As in the case of the double-sided plate problem, the numerical results indicate that solutions for the present geometry can only be generated under superalfvenic flow conditions, that is, when $\beta < 1$. The values of the α and β parameters for which the boundary conditions $g_0' = g'(\infty) = 1$ are satisfied are plotted in figures 2 and 3 as a function of β for selected ε values. The magnetic profiles obtained in the low β-value range, when $\varepsilon = 1$, conforms accurately with that inferred from equations (7) for the $\beta = 0$ case. As $\beta \to 1$, however, the g_0 parameter is seen to grow unbounded (with negative values), while the α parameter decreases to zero. A similar dependence is also observed in $\varepsilon < 1$ cases except that both parameters keep finite (non-zero) values in the $\beta = 1$ limit. The values of the α parameter in this limit increase with decreasing ε and coincide with that of the BLASIUS problem (non-conducting flows) when $\varepsilon = 0$.

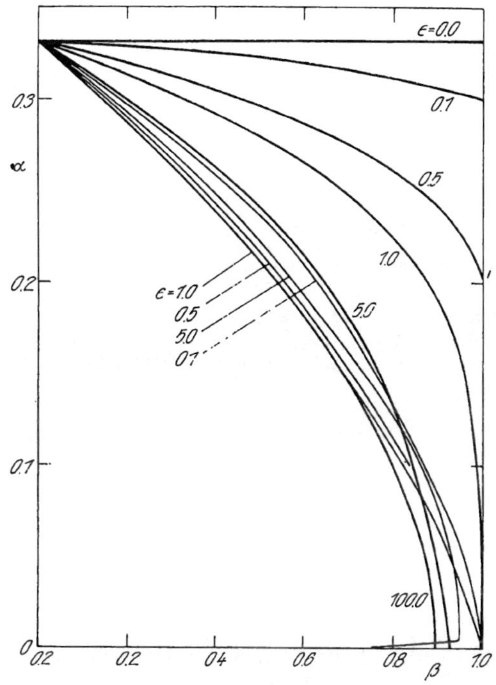

Fig. 2. Values of the skin friction parameter α as a function of β for selected values of ε. The corresponding values for the thin plate are also shown (light curves) for comparison

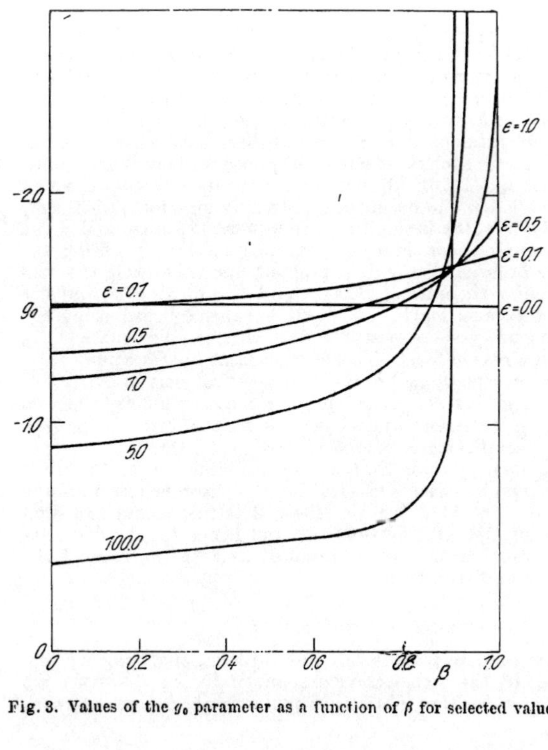

Fig. 3. Values of the g_0 parameter as a function of β for selected values of ε

In the $\varepsilon > 1$ range, on the other hand, the skin friction parameter α decreases to zero at $\beta < 1$ values so that flow conditions for which no solutions can be generated are, in the present case, also present in the superalfvenic regime. The α values obtained in this range are seen to approach rapidly those for the $\varepsilon = \infty$ case which defines the minimum skin friction values possible for the present geometry (the values given in fig. 2 for $\varepsilon = 100$ remain unaltered up to the second decimal as $\varepsilon \to \infty$). No dual solutions similar to those reported by STEWARTSON and WILSON (1964) for the double-sided plate problem are encountered in the present case as $\alpha \to 0$. The maximum β value for which α becomes identically zero decreases from $\beta = 1$ in the $\varepsilon = 1$ case to $\beta^* \simeq 0.91$ as $\varepsilon \to \infty$. The non existence of solutions in the $\beta^* < \beta < 1$ range does not appear to be related (as in the subalfvenic regime) to the conditions in the potential flow but is determined by local magnetic effects which become particularly strong with increasing ε. Most important is an effective enhancement of the viscous stress within the fluid to values larger than those forced by the non-slip condition at the wall. Such an enhancement (which can be appreciated from the second derivative of 7a) is seen to occur for all non zero values of the β parameter and increases with the g_0 parameter which in turn is seen to grow unbounded as $\alpha \to 0$. The implication of taking increasingly larger negative values of the g_0 parameter is that the resulting magnetic geometry yields an electric current system of ever increasing spatial extent away from the plate. This in turn implies the thickening of the velocity boundary layer and an ensuing large displacement velocity component $v_y(\infty)$, thus violating the basic geometric restrictions of the boundary layer approximation.

At small values of the ε and β parameters, on the other hand, the induced magnetic field produces only slight changes in the transverse component of the magnetic field above the plate. This quantity is, therefore, approximately conserved across the boundary layer and makes the magnetic field at infinity to become aligned to the (displaced) velocity vector in the free-stream flow. As indicated before, this ensures that no electric currents exist in the potential flow region.

References

1 CABANNES, H., Theoretical Magnetofluiddynamics, Academic Press, New York, 1970.
2 GREENSPAN, H.; CARRIER, G. F., The Magnetohydrodynamic Flow Past a Flat Plate, JFM 6, 77, 1959.
3 MEKSYN, D., New Methods in Laminar Boundary Layer Theory, Pergamon Press, Oxford, England, 1961.
4 ROBERTS, P. R., An Introduction to Magnetohydrodynamics, Longmans, 1967.
5 STEWARTSON, K.; WILSON, D. H., Dual Solutions of the Greenspan-Carrier Equations II, JFM 18, 337, 1964.

Eingereicht am 25. 7. 1980, revidierte Fassung am 31. 7. 1981

Address: Dr. H. PÉREZ-DE-TEJADA, Instituto de Geofisica, Universidad Nacional Autonoma de México, México 20, D. F. and Dr. M. S. UBEROI, Aerospace Engineering Sciences Dept., University of Colorado, Boulder, Colorado, USA

ZAMM 62, 417 – 418 (1982)

T. M. ATANACKOVIĆ / DJ. S. DJUKIĆ

A Variational Principle for a Diffusion Problem

For mass diffusion (heat conduction) in a solid with positive concentration (temperature) dependent sources the following boundary value problem arises [1]

$$\varphi'' + \exp \varphi = 0\,, \qquad 0 < x < 1\,, \tag{1}$$

$$\varphi(0) = \varphi(1) = 0\,. \tag{2}$$

The problem (1), (2) has been treated by many different methods. In [2] an approximate solution to (1), (2) was obtained by a variational principle of Hamilton type, and an error estimate procedure based on the value of the square residuale was used to estimate the error of this solution.

In this note we will formulate a variational principle for (1), (2) and use it to obtain an approximate solution. After that, on the basis of the variational principle an error bound will be derived. Both the variational principle and the error estimate procedure are new. The problem (1), (2) belongs to a class of boundary value problems for which a variational theory was developed in [3]. From [3] we conclude that the functional

$$I = \int_0^1 \{\Phi'^2 - \exp \Phi + \Phi'' \ln(-\Phi'') - \Phi''\} \, dx \tag{3}$$

attains a maximum on the solution of (1), (2). Let $\Phi = \varphi + \delta\varphi$ be an approximate solution of (1), (2) that satisfies the boundary conditions identically. Expanding (3) and using the fact [3] that $I(\varphi) = 0$, we have

$$I(\varphi) = \frac{1}{2} \int_0^1 \left\{ \frac{1}{\varphi'' + \varepsilon(\Phi'' - \varphi'')} \delta\varphi''^2 + 2\delta\varphi'^2 - \exp[\varphi + \varepsilon(\Phi - \varphi)] \, \delta\varphi^2 \right\} dx \,, \tag{4}$$

where $0 < \varepsilon < 1$. However, from [1] the bounds on φ and φ'' are

$$0 \leqq \varphi \leqq 0.142 \,, \qquad -\varphi'' \leqq \exp 0.142 \,. \tag{5}$$

Therefore, equation (4) becomes

$$-2I \geqq \int_0^1 \left\{ \frac{1}{R_{\max}} \delta\varphi''^2 - 2\delta\varphi'^2 + R_{\min} \, \delta\varphi^2 \right\} dx \,, \tag{6}$$

where $R_{\max} = \max[-\varphi'', -\Phi'']$ and $R_{\min} = \min[\exp \varphi, \exp \Phi]$. Now, using the inequality (cf. [4])

$$\int_0^1 \delta\varphi'^2 \, dx \leqq \varepsilon_1 \int_0^1 \delta\varphi''^2 \, dx + K(\varepsilon_1) \int_0^1 \delta\varphi^2 \, dx \,, \tag{7}$$

where $K(\varepsilon_1) = \dfrac{1}{\varepsilon_1} + 12$, $\varepsilon_1 > 0$, and the following two inequalities (cf. [5])

$$\int_0^1 \delta\varphi''^2 \, dx \geqq \pi^2 \int_0^1 \delta\varphi'^2 \, dx \,, \tag{8}$$

$$||\delta\varphi||_{L_\infty} = \sup_{x \in (0,1)} |\delta\varphi| \leqq \frac{1}{2} \left(\int_0^1 \delta\varphi'^2 \, dx \right)^{1/2} , \tag{9}$$

in relation (6) we obtain the *error estimate*

$$||\delta\varphi||_{L_\infty} \leqq \left[\frac{-I}{2\left\{\pi^2 \left[\dfrac{1}{R_{\max}} - \dfrac{R_{\min}\varepsilon_1^2}{1 + 12\,\varepsilon_1}\right] + \left[\dfrac{R_{\min}\varepsilon_1}{1 + 12\,\varepsilon_1} - 2\right]\right\}} \right]^{1/2} . \tag{10}$$

It is interesting to note that the error estimate (10) could be "optimized" by choosing such a value for $\varepsilon_1 > 0$ that will make the right hand side of (10) minimal.

For completeness we mention the L_2 error bound derived in [3],

$$||\delta\varphi||_{L_2} \leqq \left\{ \frac{-2I}{[G(\lambda_n^2)]_{\min}} \right\}^{1/2} , \tag{11}$$

where

$$G(\lambda_n^2) = \frac{\lambda_n^4}{R_{\max}} - 2\lambda_n^2 + R_{\min} \,, \tag{12}$$

and $\lambda_n = n\pi$, $n = 1, 2, \ldots$

We have performed calculations with the trial functions of the form

$$\Phi = \sum_{n=1}^{n=3} a_n (x - x^2)^n \,. \tag{13}$$

Substituting (13) into (3) and maximizing with respect to a_n the following values are found

Table

a_1	a_2	a_3	$I(\varphi)$
0.549	0.048	0.0129	$3.25155 \cdot 10^{-7}$

To estimate the error, note that in the present case

$$R_{\min} = 1 \,, \qquad R_{\max} = \exp 0.142 \,. \tag{14}$$

Also, we take $\varepsilon_1 = 1/2\pi^2$. The inequality (10) then gives

$$||\delta\varphi||_{L_\infty} \leqq 1.572 \cdot 10^{-4} \,, \tag{15}$$

and the inequality (11)

$$||\delta\varphi||_{L_2} \leqq 0.994 \cdot 10^{-4} \,. \tag{16}$$

From (15) and (16) we conclude that, although simple, the approximate solution (13) has remarkable accuracy.

References

1 BAILEY, P B ; SHAMPINE, L. F ; WALTMAN, P E., Nonlinear Two-Point Boundary Value Problems, Academic Press, New York 1968.
2 ARTHURS, A M , WINTHROP, D., Variational solution and error bound for a nonlinear boundary value problem in heat diffusion, Nuovo Cimento Lett 7, 150 – 152 (1973).
3 DJUKIĆ, DJ S., ATANACKOVIĆ, T M , Error bounds via a new extremum variational principle, Mean square residual and weighted mean square residual, J Math Analysis Appl 75, 203 – 218 (1980).
4 MITRINOVIĆ, D S., Analytic Inequalities, Springer, Berlin 1970.
5 SHAMPINE, L. F , Error bounds and variational methods for nonlinear boundary value problems, Numer Math. 12, 410 – 415 (1968).

Eingereicht: 23. 7. 1981

Anschrift: Dr. T. ATANACKOVIĆ, Dr. DJ. DJUKIĆ, Fakultet tehničkih nauka, 21000 Novi Sad, V. Vlahovica 3, Yugoslavia

ZAMM 62, 418 – 420 (1982)

T. R. GULATI

On Nonlinear Nondifferentiable Complex Programming Problems

1. Introduction

Duality for a nonlinear nondifferentiable complex programming problem with linear constraints was studied by BHATIA and KAUL [3] and MOND [12]. The nondifferentiability occurs in the objective function, which is the sum of a linear term and square root of a quadratic form. These results were extended by adding a quadratic form and taking a finite sum of square roots of quadratic forms in the objective function by RANI and KAUL [15], RANI [14] and JAIN and SAXENA [9]. DAS and SWARUP [6] studied duality theory by adding another quadratic form in the objective function of the primal problem of [14, 9]. The problems in [3, 15] are over special polyhedral cones and those in [12, 14, 9, 6] are over arbitrary polyhedral cones. The duals in all these papers are differentiable. Several authors have obtained symmetric duals for such problems with special nonlinear constraints.

BHATIA and KAUL [3] proved a converse duality theorem using duality in homogeneous programming, for which they assumed a hypothesis similar to EISENBERG [7]. RANI and KAUL [15] obtained a converse duality theorem by linearizing the objective function of the dual problem and then using the converse duality theorem of BHATIA and KAUL [3]. The converse duality theorems in [12, 14, 9, 6] were established using the Kuhn-Tucker type necessary conditions in complex space [1, 2], which requires the Kuhn-Tucker constraint qualification.

In the present paper we consider the problems of RANI [14] and JAIN and SAXENA [9] and prove a converse duality theorem under three different hypotheses, using the Fritz John type necessary conditions in complex space [4, 5]. One of our hypotheses is weaker than that of BHATIA and KAUL [3], while another gives a strict converse duality theorem (see footnote on page 117 in [11]). We also show that the primal and dual problems of DAS and SWARUP [6], and hence the duality relations between them, can be obtained as a special case of the problems considered here.

2. Notations and definitions

Let $\mathbb{C}^n(\mathbb{R}^n)$ denote respectively the n-dimensional complex vector space, $\mathbb{R}^n_+$ the nonnegative orthant of $\mathbb{R}^n$ and $\mathbb{C}^{m \times n}$ the set of all $m \times n$ complex matrices. Superscript H denotes the conjugate transpose. $S \subset \mathbb{C}^n$ is a polyhedral cone if it is the finiet intersection of closed half spaces in $\mathbb{C}^n$, each containing 0 in its boundary. The polar S^* of a nonempty set $S \subset \mathbb{C}^n$ is defined by

$$S^* = \{y \in \mathbb{C}^n : x \in S \Rightarrow \operatorname{Re}(y^H x) \geqq 0\} .$$

3. Converse duality

Rani [14] and Jain and Saxena [9] discussed duality theorems for the following pair of primal and dual problems:

Primal problem (P1):

$$\text{Minimize} \quad f(u) = \operatorname{Re}\left[p^H u + \tfrac{1}{2} u^H C u + \sum_{i=1}^{r} (u^H Q^i u)^{1/2}\right]$$

$$\text{subject to} \quad Au - b \in S , \qquad u \in T ,$$

Dual problem (D1):

$$\text{Maximize} \quad g(x, y) = \operatorname{Re}\left[b^H y - \tfrac{1}{2} x^H C x\right]$$

$$\text{subject to} \quad -A^H y + Cx + \sum_{i=1}^{r} Q^i w^i + p \in T^* ,$$

$$w^{iH} Q^i w^i \leqq 1 , \qquad i = 1, 2, \dots, r; \qquad y \in S^* ,$$

where $A \in \mathbb{C}^{m\times n}$, $b \in \mathbb{C}^m$, $p \in \mathbb{C}^n$; C and Q^i, $i = 1, 2, \dots, r$, are $n \times n$ positive semidefinite Hermitian matrices; $S \subset \mathbb{C}^m$ and $T \subset \mathbb{C}^n$ are polyhedral cones.

Rani and Kaul [15] studied the above problems over special polyhedral cones S and T defined by

$$S = \{z \in \mathbb{C}^m : |\arg z| \leqq \alpha\} , \qquad T = \{v \in \mathbb{C}^n : |\arg v| \leqq \beta\}$$

for given $\alpha \in \mathbb{R}^m_+$, $\beta \in \mathbb{R}^n_+$, $\alpha \leqq \pi/2$, $\beta \leqq \pi/2$ ($\pi/2$ being a vector of appropriate dimension with $\pi/2$ in each position). They obtained a converse duality theorem using (implicitly) duality in homogeneous programming [7, 16] under the hypothesis:

$$Au \in S , \qquad u \in T , \qquad f(u) \leqq 0 \Rightarrow u = 0 .$$

Rani [14] proved a converse duality theorem assuming the Kuhn-Tucker constraint qualification for the dual problem in order to apply the Kuhn-Tucker type necessary conditions [1, 2]. Jain and Saxena [9] provided two proofs of their converse duality theorem. In the first proof (Theorem 4.1 (d)), they wrote the dual problem in the primal format and then applied their direct duality theorem. However, since there is an additional nonlinear constraint, the direct duality theorem does not apply here. Their second proof (Theorem 5.1) is exactly the same as that of Rani [14], but they did not assume any constraint qualification. Mond [13] observed that the corresponding footnote is not correct. Below we prove a converse duality theorem under three different hypotheses using the Fritz John type necessary conditions [4, 5]. We shall need the following lemma:

Lemma 1 [12]: *Let $Q \in \mathbb{C}^{n\times n}$ be a positive semidefinite Hermitian matrix. Then*

$$\operatorname{Re}(\eta^H Q w) \leqq (\eta^H Q \eta)^{1/2} (w^H Q w)^{1/2} .$$

Equality holds if, for some $\lambda \geqq 0$, $Q\eta = \lambda Q w$.

Theorem 1: *Let (w_0^i, x_0, y_0), $i = 1, 2, \dots, r$, be an optimal solution of the dual problem and let any one of the following three hypotheses be satisfied:*

(H1): *the matrix C is nonsingular;*

(H2): *for some i the matrix Q^i is nonsingular and $w_0^{iH} Q^i w_0^i < 1$;*

(H3): $Au \in S , \quad u \in T , \quad f(u) = 0 \Rightarrow u = 0 .$

Then there exist an optimal solution u_0 of the primal problem and real numbers $\alpha_i \geqq 0$ such that $Cu_0 = Cx_0$, $Q^i u_0 = \alpha_i Q^i w_0^i$ and $(u_0) = g(x_0, y_0)$.

Proof: Since (w_0^i, x_0, y_0), $i = 1, 2, \dots, r$, is an optimal solution of the dual problem, by the Fritz John type necessary conditions in complex space [4, 5] there exist $\tau \in \mathbb{R}_+$, $\xi \in S$, $\eta \in T$ and $\lambda_i \in \mathbb{R}_+$, $i = 1, 2, \dots, r$, satisfying

$$A\eta - \tau b = \xi \in S , \qquad C\eta = \tau C x_0 , \tag{1), (2}$$

$$Q^i \eta - \lambda_i Q^i w_0^i , \qquad i - 1, 2, \dots, r , \tag{3}$$

$$\operatorname{Re}\left[\left(-y_0^H A + x_0^H C + \sum_{i=1}^{r} w_0^{iH} Q^i + p^H\right) \eta\right] = 0 , \tag{4}$$

$$(w_0^{iH} Q^i w_0^i - 1)\, \lambda_i = 0 , \qquad i = 1, 2, \dots, r , \tag{5}$$

$$\operatorname{Re}[y_0^H \xi] = 0 , \qquad (\tau, \lambda, \xi, \eta) \neq 0 , \tag{6), (7}$$

where $\lambda = (\lambda_1, \lambda_2, \dots, \lambda_r)$.

Suppose $\tau = 0$. We first show that each hypothesis implies $\eta = 0$.

(H1): From (2) $C\eta = 0$. Therefore $\eta = 0$.

(H2): Let Q^k be nonsingular and $w_0^{kH} Q^k w_0^k < 1$. Therefore, from (5), $\lambda_k = 0$, which with (3) gives $\eta = 0$.

(H3): This hypothesis needs a longer proof. From (1) and (6),

$$\operatorname{Re}(y_0^H A\eta) = \operatorname{Re}(y_0^H \xi) = 0 . \tag{8}$$

From (2),

$$C\eta = 0 . \tag{9}$$

Since (3) is the condition of equality in Lemma 1,

$$\operatorname{Re}(\eta^H Q^i w_0^i) = (\eta^H Q^i \eta)^{1/2} (w_0^{iH} Q^i w_0^i)^{1/2} , \qquad i = 1, 2, \dots, r . \tag{10}$$

From (5), for each i, either $\lambda_i = 0$ or $w_0^{iH} Q^i w_0^i = 1$. In either case, from (3) and (10) we get

$$\operatorname{Re}(\eta^H Q^i w_0^i) = (\eta^H Q^i \eta)^{1/2} , \qquad i = 1, 2, \dots, r . \tag{11}$$

Now from (9), (11) and (4) we obtain

$$f(\eta) = \operatorname{Re}\left[p^H \eta + \tfrac{1}{2} \eta^H C \eta + \sum_{i=1}^{r} (\eta^H Q_\eta^i)^{1/2}\right] =$$

$$= \operatorname{Re}\left[p^H \eta + \sum_{i=1}^{r} (\eta^H Q^i w_0^i)\right] = \operatorname{Re}\left[y_0^H A\eta - x_0^H C_\eta^i\right] = 0 .$$

Therefore we have $A\eta \in S$, $\eta \in T$, $f(\eta) = 0$. This implies $\eta = 0$.

Thus each hypothesis gives $\eta = 0$. Now from (1) $\xi = 0$, and from (3) and (5) $\lambda_i = 0$, $i = 1, 2, \dots, r$. Therefore $(\tau, \lambda, \xi, \eta) = 0$, a contradiction to (7). Hence $\tau > 0$; and from (1) and $\eta \in T$ we obtain that $u_0 = \eta/\tau$ is a feasible solution for the primal problem. Substituting $\eta = \tau u_0$ in (2) and (3) we get

$$Cu_0 = Cx_0 \tag{12}$$

and

$$Q^i u_0 = \alpha_i Q^i w_0^i , \qquad i = 1, 2, \dots, r , \tag{13}$$

where $\alpha_i = \lambda_i/\tau$. Also, as above,

$$f(u_0) = \operatorname{Re}\left[p^H u_0 + \tfrac{1}{2} u_0^H C u_0 + \sum_{i=1}^{r} (u_0^H Q^i u_0)^{1/2}\right] =$$

$$= \operatorname{Re}\left[p^H u_0 + \tfrac{1}{2} u_0^H C x_0 + \sum_{i=1}^{r} (u_0^H Q^i w_0^i)\right] =$$

$$= \operatorname{Re}\left[y_0^H A u_0 - \tfrac{1}{2} u_0^H C x_0\right] = \operatorname{Re}\left[(y_0^H b + y_0^H \xi/\tau - \tfrac{1}{2} x_0^H C u_0\right] =$$

$$= \operatorname{Re}\left[b^H y_0 - \tfrac{1}{2} x_0^H C x_0\right] = g(x_0, y_0) .$$

Hence, by the weak duality theorem [14, 9], u_0 is an optimal solution of the primal problem. This completes the proof.

Remark 1: If the matrix C is nonsingular (Hypothesis (H1)), then from (12) $u_0 = x_0$. Therefore, in this case, Theorem 1 becomes a strict converse duality theorem. Similarly, if for some $i = k$, the matrix Q^k is nonsingular (Hypothesis (H2)), then we get the relation $u_0 = \alpha_k w_0^k$ between the optimal solutions of the primal and dual problems.

Remark 2: Similarly to [8], it can be shown that the proof of the converse duality theorem given by Bhatia and Kaul [3] can be modified to hold under the weaker hypothesis (H3), but then it becomes longer and complicated.

4. The problems of Das and Swarup

In this section we show that the following problems studied by Das and Swarup [6] can be obtained as a particular case of the problems considered in the last section.

Primal problem (P2):

$$\text{Minimize} \quad F(u, v) = \operatorname{Re}\left[p^H u + \tfrac{1}{2} u^H C u + \tfrac{1}{2} v^H D v + \sum_{i=1}^{r} (u^H Q^i u)^{1/2}\right]$$

$$\text{subject to} \quad Au + Bv - b \in S , \qquad u \in T ,$$

Dual problem (D2):

$$\text{Maximize} \quad G(x, y, z) = \operatorname{Re}\left[b^H y - \tfrac{1}{2} x^H C x - \tfrac{1}{2} z^H D z\right]$$

$$\text{subject to} \quad -A^H y + Cx + \sum_{i=1}^{r} Q^i w^i + p \in T^* ,$$

$$w^{iH} Q^i w^i \leqq 1 , \qquad i = 1, 2, \dots, r, \qquad B^H y = Dz , \qquad y \in S^* ,$$

where $B \in \mathbb{C}^{m \times m}$ and D is an $m \times m$ positive semidefinite Hermitian matrix. All other symbols are as defined in Section 3. The dual problem in [6] seems to contain some errors which are corrected in (D2).

Remark 3: MAHAJAN and VARTAK [10] have observed that the order of the matrix D need not be equal to the number of rows in A.

The problem (P2) can be written as:
Minimize

$$F(u, v) = \mathrm{Re}\left[\begin{pmatrix} p \\ 0 \end{pmatrix}^{\mathsf{H}} \begin{pmatrix} u \\ v \end{pmatrix} + \frac{1}{2} \begin{pmatrix} u \\ v \end{pmatrix}^{\mathsf{H}} \begin{pmatrix} C & 0 \\ 0 & D \end{pmatrix} \begin{pmatrix} u \\ v \end{pmatrix} + \right.$$

$$\left. + \sum_{i=1}^{r} \left\{ \begin{pmatrix} u \\ v \end{pmatrix}^{\mathsf{H}} \begin{pmatrix} Q^i & 0 \\ 0 & 0 \end{pmatrix} \begin{pmatrix} u \\ v \end{pmatrix} \right\}^{1/2} \right]$$

$$\text{subject to} \quad (A, B) \begin{pmatrix} u \\ v \end{pmatrix} - b \in S, \qquad \begin{pmatrix} u \\ v \end{pmatrix} \in T \times C^m,$$

where 0 appearing at several places in the objective function is a vector or matrix of appropriate dimension with all elements zero.

The above problem is in the form of (P1), and from (D1) its dual is:

$$\text{Maximize} \quad G(x, y, z) = \mathrm{Re}\left[b^{\mathsf{H}} y - \frac{1}{2} \begin{pmatrix} x \\ z \end{pmatrix}^{\mathsf{H}} \begin{pmatrix} C & 0 \\ 0 & D \end{pmatrix} \begin{pmatrix} x \\ z \end{pmatrix}\right]$$

subject to

$$-(A, B)^{\mathsf{H}} y + \begin{pmatrix} C & 0 \\ 0 & D \end{pmatrix} \begin{pmatrix} x \\ z \end{pmatrix} + \sum_{i=1}^{r} \begin{pmatrix} Q^i & 0 \\ 0 & 0 \end{pmatrix} \begin{pmatrix} w^i \\ t^i \end{pmatrix} +$$

$$+ \begin{pmatrix} p \\ 0 \end{pmatrix} \in T^* \times \{0\},$$

$$\begin{pmatrix} w^i \\ t^i \end{pmatrix}^{\mathsf{H}} \begin{pmatrix} Q^i & 0 \\ 0 & 0 \end{pmatrix} \begin{pmatrix} w^i \\ t^i \end{pmatrix} \leqq 1, \qquad i = 1, 2, \dots, r, \qquad y \in S^*.$$

On simplification this reduces to:

$$\text{Maximize} \quad G(x, y, z) = \mathrm{Re}\,[b^{\mathsf{H}} y - \tfrac{1}{2} x^{\mathsf{H}} C x - \tfrac{1}{2} z^{\mathsf{H}} D z]$$

$$\text{subject to} \quad -A^{\mathsf{H}} y + Cx + \sum_{i=1}^{r} Q^i w^i + p \in T^*,$$

$$B^{\mathsf{H}} y = Dz, \qquad w^{i\mathsf{H}} Q^i w^i \leqq 1, \qquad i = 1, 2, \dots, r, \qquad y \in S^*$$

which is the problem (D2).

In view of the above observation the following duality relations between problems (P2) and (D2) are obtained by using the corresponding results in [14], [9] and the last section. For brevity we shall write

$$(w_0, x_0, y_0, z_0) \quad \text{for} \quad (w_0^i, x_0, y_0, z_0), \qquad i = 1, 2, \dots, r.$$

Theorem 2: *If* (u_0, v_0) *and* (w_0, x_0, y_0, z_0) *are feasible solutions of the problems* (P2) *and* (D2) *respectively, then*

$$F(u_0, v_0) \geqq G(x_0, y_0, z_0).$$

Theorem 3: *If the problem* (P2) *is infeasible, then* (D2) *is infeasible or unbounded.*

Theorem 4: *If* (P2) *has an optimal solution* (u_0, v_0), *then their exists an optimal solution* (w_0, x_0, y_0, z_0) *of* (D2) *such that* $x_0 = u_0$ *and* $z_0 = v_0$, *and the optimal values of* (P2) *and* (D2) *are equal.*

Theorem 5: *Let* (D2) *have an optimal solution* (w_0, x_0, y_0, z_0) *and let any one of the following four hypotheses be satisfied:*

(H1): *The matrices C and D are nonsingular.*

(H2): *D and Q^i, for some i, are nonsingular and* $w_0^{i\mathsf{H}} Q^i w_0^i < 1$.

(H3): $Au + Bv \in S, \quad u \in T, \quad F(u, v) = 0 \Rightarrow u = 0, \; v = 0.$

(H4): (D2) *satisfies the Kuhn-Tucker constraint qualification* [1, 2].

Then there exist an optimal solution (u_0, v_0) *of* (P2) *and real numbers* β_i, $i = 1, 2, \dots, r$, *such that* $Cu_0 = Cx_0$, $Dv_0 = Dz_0$ *and* $Q^i u_0 = \beta_i Q^i w_0^i$, $i = 1, 2, \dots, r$. *Also, the optimal values of* (P2) *and* (D2) *are equal.*

Theorem 6: *Let* (u_0, v_0) *and* $(w_0, x_0 = u_0, y_0, z_0 = v_0)$ *be optimal solutions of problems* (P2) *and* (D2) *respectively.*

$$\mathrm{Re}\,[u_0^{\mathsf{H}}(-A^{\mathsf{H}} y_0 + Cu_0 + \sum_{i=1}^{r} Q^i w_0^i + p)] = 0$$

and

$$\mathrm{Re}\,[y_0^{\mathsf{H}}(Au_0 + Bv_0 - b)] = 0.$$

Remark 4: DAS and SWARUP [6] used the Kuhn-Tucker type necessary condition [1, 2] to prove the converse duality theorem without stating any constraint qualification.

Acknowledgement

The author is thankful to Dr. B. D. CRAVEN for his valuable comments on a preliminary version of this paper.

References

1 ABRAMS, R. A.; BEN-ISRAEL, A., Nonlinear programming in complex space: Necessary conditions, SIAM J. Control **9**, 606–620 (1971).
2 ABRAMS, R. A.; BEN-ISRAEL, A., Complex mathematical programming, in Developments in Operations Research (B. AVI-ITZHAK, ed.), Gordon and Breach, New York 1971.
3 BHATIA, D., KAUL, R. N., Nonlinear programming in complex space, J. Math. Analysis Appl. **28**, 144–152 (1969).
4 CRAVEN, B. D., MOND, B., A Fritz John theorem in complex space, Bull. Austral. Math. Soc. **8**, 215–220 (1973).
5 CRAVEN, B. D.; MOND, B., Real and complex Fritz John theorems, J. Math Analysis Appl. **44**, 773–778 (1973).
6 DAS, C.; SWARUP, K., A class of nonlinear nondifferentiable complex programming, ZAMM **57**, 481–484 (1977).
7 EISENBERG, E., Duality in homogeneous programming, Proc. Amer. Math. Soc. **12**, 783–787 (1961).
8 GULATI, T. R., On converse duality for a nondifferentiable program, Bull. Austral. Math. Soc. **25** 71–80 (1982).
9 JAIN, O. P.; SAXENA, P. C., A duality theorem for a special class of programming problems in complex space, J. Optimization Theory Appl. **16**, 207–220 (1975).
10 MAHAJAN, D. G.; VARTAK, M. N., Generalization of some nonlinear programming problems in complex space, J. Indian Statist. Assoc. **14**, 35–51 (1976).
11 MANGASARIAN, O. L., Nonlinear Programming, McGraw-Hill Book Co., New York 1969.
12 MOND, B., Nonlinear nondifferentiable programming in complex space, in Nonlinear Programming (J. ROSEN, O. L. MANGASARIAN, and K. RITTER, eds.), Academic Press, New York 1970, 385–400.
13 MOND, B., On converse duality for a nondifferentiable programming problem, Operations Research **25**, 713–715 (1976).
14 RANI, O., A duality theorem for complex nonlinear programming, Opsearch **10**, 14–23 (1973).
15 RANI, O.; KAUL, R. N., Nonlinear programming in complex space, J. Math. Analysis Appl. **43**, 1–14 (1973).
16 SMILEY, M. F., Duality in complex homogeneous programming, J. Math. Analysis Appl. **40**, 153–158 (1972).

Eingereicht: 27. 7. 1981

Anschrift: Dr. T. R. GULATI, Department of Mathematics, University of Roorkee, Roorkee 247672, India and Department of Mathematics, University of Melbourne, Parkville, Victoria 3052, Australia

BUCHBESPRECHUNGEN

Nemat-Nasser, S. (ed.), Mechanics Today, vol. 6. Oxford-Frankfurt, Pergamon Press 1981. XXII, 204 S., US $ 50.00. ISBN 0-08-024749-0

Der vorliegende Band 6 der von dem international bekannten Festkörpermechanik- und Bruchmechanik-Spezialisten NEMAT-NASSER, USA, herausgegebenen Reihe "Mechanics Today" reiht sich würdig ein in die bisher darin publizierten ausgezeichneten Darstellungen zur Mechanik. Besonders hervorzuheben an dem vorliegenden Sammelband ist wiederum der Versuch, eine direkte Synthese mathematischer, physikalischer und ingenieurwissenschaftlicher Gesichtspunkte zu erreichen. Das Buch enthält 4 Kapitel. Die drei ersteren bestehen aus je einer ausführlichen Arbeit mit Übersichtscharakter zu einigen sehr aktuellen Gebieten der theoretischen und der angewandten Mechanik, während Kapitel IV eine kleinere Ergänzung von F. ERDOGAN zu einer von ihm in Band 4 dieser Reihe erschienenen Arbeit über gemischte Randwertprobleme der Mechanik ist. Kapitel I aus der Feder des Amerikaners J. J. MC COY, beschäftigt sich mit einem allgemeinen Zugang zur Kontinuumsmechanik auf statistischer Grundlage, wobei die Makro-Mikrorelationen ausführlich behandelt werden. So wichtige Fragen wie „effektive" elastische Werkstoffkonstanten fehlen dabei nicht. Auch die gegenwärtig sehr aktuellen Fragen der nichtlokalen Feldtheorie der Festkörpermechanik werden kurz berührt. Wertvoll ist auch die ausführliche Quellenangabe mit ca. 100 Titeln. Kapitel II wurde von M. J. SEWELL, Großbritannien, verfaßt. Hier stehen sowohl Anwendungen in der Mechanik als auch in der Physik eines Gebietes zur Diskussion, das als Bindeglied zwischen Katastrophentheorie und Variationsrechnung einzuordnen ist. Die Arbeit ist mathematisch verständlich geschrieben und auch durch zahlreiche Abbildungen bereichert, was dieses i. a. anspruchsvolle Fachgebiet auch dem Nichtexperten näherbringen könnte. Auch hier ist wiederum das gute Quellenverzeichnis hervorzuheben. In Kapitel III, das von J. ZARKA und J. CASIER, Frankreich, verfaßt wurde, steht das Grenzverhalten elastisch-plastischer Konstruktionen unter zyklischer Belastung im Mittelpunkt. Dabei wurde der Schwerpunkt klar auf eine ausführliche Diskussion von Beispielen gelegt.

Auch der vorliegende Band der Reihe "Mechanics Today" dürfte sowohl für Fachkollegen der Festkörpermechanik, der Physik und der Mathematik von besonderem Interesse sein.

Karl-Marx-Stadt B. MICHEL

Hult, J. / Lemaitre, J. (eds.), Physical Non-Linearities in Structural Analysis. IUTAM-Symposium Senlis, France, 1980. Berlin-Heidelberg-New York, Springer-Verlag 1981. XI, 287 S., 109 Abb., DM 64,–. US $ 30.50. ISBN 3-540-10544-1

Der Band enthält 39 Beiträge. Sie behandeln überwiegend plastisches oder viskoplastisches Materialverhalten, oft in Verbindung mit Fragen von Schadens- oder Bruchkonzepten sowie Anwendungen in der Strukturmechanik. Auch Probleme der numerischen Realisierung werden verschiedentlich angeschnitten.

Berlin Redaktion

Truckenbrodt, E., Fluidmechanik, Bd. 2: Elementare Strömungsvorgänge dichteveränderlicher Fluide sowie Potential- und Grenzschichtströmungen, 2., völlig neubearb. u. erw. Aufl., Berlin-Heidelberg-New York, Springer-Verlag 1980. XVI, 426 S., 177 Abb., DM 136,–. US $ 80.30. ISBN 3-540-10135-7

Das vorliegende Buch bildet den 2. Teil der „Fluidmechanik" von E. TRUCKENBRODT, die als neu bearbeitete und erweiterte zweite Auflage der „Strömungsmechanik" erscheint (s. Buchbesprechung des 1. Teils in ZAMM 61, S. 275). Dieser zweite Band umfaßt die Kapitel 4 bis 6. In Kapitel 4 werden elementare Strömungsvorgänge dichteveränderlicher Fluide behandelt. Wie auch in Kapitel 3 wird das Verhalten im Ruhezustand einbezogen, die Anwendungen auf einfach zu überschauende elementare Strömungsvorgänge stehen im Mittelpunkt. Neben der Aerostatik werden die Stromfadentheorie und Strömungen in Rohrleitungen bei dichteveränderlichen Fluiden und die Umlenkung stationärer ebener Überschallströmungen durch Wellen und Stöße dargestellt. Kapitel 5 faßt die drehungsfreien und -behafteten Strömungen zusammen. Es enthält Begriffe und Grundgesetze, die Darstellung der drehungsfreien reibungslosen Strömungen dichtebeständiger und dichteveränderlicher Fluide sowie drehungsbehaftete reibungslose Strömungen (Potentialwirbelströmungen) mit Anwendungen in der Tragflügeltheorie. Auf verwandte Probleme der Potentialtheorie, wie Strömungen mit freien Stromlinien, schleichende Strömungen und Sickerströmungen, wird ebenfalls eingegangen. Das Kapitel 6 bringt eine Einführung in die Grenzschichttheorie. Nach Begriffen und Grundgleichungen für Strömungs- und Temperaturgrenzschichten werden laminare und turbulente Grenzschichtströmungen an festen Wänden und ohne feste Begrenzung behandelt. Die Übersicht bleibt auf ebene Grenzschichten beschränkt, Hinweise auf vielfältige Anwendungsgebiete werden gegeben.

Neben den Literaturhinweisen an den Enden der Kapitel ist dem Werk eine Bibliographie beigefügt, untergliedert nach den Sachgebieten Grundlagen, Strömungen dichteveränderlicher Fluide, Grenzschichtströmungen und Angewandte Fluidmechanik.

Das Werk vermittelt die Grundlagen der Fluidmechanik in selten erreichter Anschaulichkeit, Eindringlichkeit und Vollständigkeit bei engem Bezug zu den vielfältigen Anwendungsproblemen dieses Wissensgebietes. Es kann daher allen Studierenden der naturwissenschaftlich-technischen Richtungen und den praktisch tätigen Ingenieuren und Naturwissenschaftlern, die mit diesem Fachgebiet direkt oder indirekt verbunden sind, wärmstens empfohlen werden.

Berlin H.-E. PETERS

Hirschel, E. H. / Kordulla, W., Shear Flow in Surface-Oriented Coordinates. Notes on Numerical Fluid Mechanics 4. Braunschweig-Wiesbaden, Vieweg 1981. X, 266 S., ISBN 3-528-08078-7

Die Autoren haben sich die Aufgabe gestellt, auf der Grundlage eigener langjähriger Arbeiten die beschreibenden Gleichungen für dreidimensionale Scherströmungen in der Nähe nahezu beliebig geformter Wände darzustellen und zu diskutieren. Dieses Vorhaben ist zu begrüßen, weil reale Strömungsvorgänge stets mehr oder weniger derartigen Bedingungen entsprechen und die Leistungsfähigkeit der modernen Rechentechnik die numerische Lösung so allgemeiner Fragestellungen in zunehmendem Maße ermöglicht. Auch sind zuverlässige Einschätzungen von – praktisch noch häufig vorgenommenen – Vereinfachungen oft nur unter Verwendung umfassend formulierter und gelöster Aufgabenstellungen möglich.

In Anlehnung an die bei Problemen der Luftfahrt gegebenen Bedingungen setzen die Autoren ein kompressibles Newtonsches Medium mit den thermischen Eigenschaften eines idealen Gases voraus. Die Gleichungen werden für dreidimensionale laminare und – mit Ausnahme einiger fundamentaler Zusammenhänge – stationäre Strömungen formuliert. Über die Transporteigenschaften des Mediums kann nach den Vorstellungen der Autoren die Wirkung der Strömungsturbulenz einbezogen werden. Dies gilt allerdings ohne Modifikation für die angegebenen Beziehungen nur bei Verwendung von Turbulenzmodellen, die auf skalare Transportkoeffizienten vom Gradiententyp führen. Für wandnahe dreidimensionale turbulente Scherschichten trifft das nur bedingt zu, weil u. a. aus Messungen ermittelte Wirbelzähigkeiten eine gewisse Richtungsabhängigkeit aufweisen. Auf weitere Fragen der Turbulenzmodellierung gehen die Autoren nicht ein.

Zur Darstellung der Grundgleichungen, also der Kontinuitätsgleichung, der Bewegungsgleichungen und der Energiegleichung einschließlich der Bedingungen an den Begrenzungswänden, werden oberflächenorientierte, lokal monokline Koordinaten benutzt. Diese sind dadurch charakterisiert, daß die Begrenzungswand mit einem i. a. schiefwinkligen und krummlinigen zweidimensionalen Koordinatennetz überzogen wird und die dritte Koordinate geradlinig ist und lokal senkrecht auf die Wand trifft. Mit Recht vermerken die Autoren die begrenzte Anwendbarkeit dieser Koordinaten bei stärker – im Vergleich zur Scherschichtdicke – konkav gekrümmten Oberflächen.

Die Autoren behandeln in 13 Abschnitten und 5 Anhängen u. a. die Grenzschichtgleichungen erster Ordnung, die Darstellung integraler Grenzschichtparameter, Fragen der Anfangsbedingungen für Grenzschichtrechnungen, Grenzschichtgleichungen von höherer Ordnung, Einflüsse der Oberflächenkrümmung, Beispiele von oberflächenorientierten Koordinatensystemen, die Navier-Stokesschen Gleichungen in der Vereinfachung für „dünne Schichten", Probleme der dreidimensionalen Grenzschichtablösung und Staubereiche sowie Ähnlichkeitstransformationen für dreidimensionale Grenzschichten. Für ein vertieftes Eindringen in spezielle Fragestellungen werden 133 Literaturhinweise, zumeist neueren Datums, gegeben. Ein umfangreiches Sachregister und eine ausführliche Aufstellung der benutzten Symbole erleichtern den Umgang mit dem gebotenen Stoff. Das vorliegende Buch wird zweifellos von all den Fachkollegen mit Gewinn genutzt werden können, die sich mit der theoretischen oder experimentellen Untersuchung dreidimensionaler Scherschichten unter den angegebenen Bedingungen zu befassen haben.

Berlin M. Hoffmeister

Lewis, R. W. / Morgan, K. / Zienkiewicz, O. C. (eds.), Numerical Methods in Heat Transfer. Chichester-New York-Brisbane-Toronto, John Wiley & Sons 1981. XV, 536 S., £ 28.00. ISBN 0-471-27803-3

Das Buch berichtet über die im Juli 1979 in Swansea durchgeführte internationale Konferenz zu numerischen Methoden bei der Lösung von Wärmeleitungsproblemen. Es werden 24 Beiträge bekannter Autoren wie O. C. Zienkiewicz, C. A. Brebbia, J. Crank, D. B. Spalding vorgestellt, die schwerpunktmäßig auf die Anwendung moderner numerischer Methoden bei konkreten Wärmeleitungsproblemen orientiert sind. Eine theoretisch – mathematische Fundierung dieser Methoden und Verfahren ist nicht beabsichtigt.

Methodisch befassen sich 7 Kapitel mit der Finiten Elemente-Methode, darunter eine historische Übersicht zur Entwicklung der FEM von O. C. Zienkiewicz. 10 Kapitel verwenden Differenzenverfahren und 3 Kapitel sind der Einführung und Anwendung der sehr entwicklungsfähigen Rand-Elemente-Methode (Boundary Element Method (BEM), Boundary Integral Equation Method (BIEM)) gewidmet. Hervorzuheben ist die grundlegende Einführung in die BEM von C. A. Brebbia. Darüberhinaus gibt es einen Beitrag von M. A. Hogge, der Vor- und Nachteile von Ein- und Mehrschrittverfahren diskutiert, einen Beitrag von M. D. Mikhailov und M. A. Aladjem, die eine Softwareentwicklung zur automatischen Lösung von Rand- und Anfangswertaufgaben für gewöhnliche und partielle Differentialgleichungssysteme vorstellen, und zwei Beiträge von L. Imre und H. Moen, die sich mit thermalen Problemen bei komplexen industriellen Prozessen beschäftigen.

Schwerpunkte der numerischen Behandlung sind Berechnungsverfahren für zwei- und teilweise auch dreidimensionale kompliziert berandete Gebiete. Hervorzuheben sind hierzu die BE-Methoden zu thermoelastischen und -viskoelastischen Problemen von A. Chaudouet, G. Loubignac und M. Predeleanu sowie die vielfältigen Untersuchungen zur Geometrievariabilität von Differenzenverfahren. Die Anwendung krummliniger Koordinaten zur Berechnung nichtisothermer Umströmprobleme von D. Dutoya und P. Michard, die Beschreibung des Wärmeaustausches an freien Oberflächen und Phasenflächen mit bewegten Netzen von J. Rank u. a. und die Behandlung dreidimensionaler Rezirkulationsströmungen in Brennkammern von D. B. Spalding und M. A. Serag-Eldin stellen bemerkenswerte Beispiele für den letztgenannten Problemkreis dar. Weitere numerische Probleme wurden mit der Einbeziehung von Turbulenzaspekten (k-ε-Modelle, turbulente Diffusion) und mit der Behandlung von konvektiv-dominanten Energietransportprozessen aufgegriffen.

Von den 24 Beiträgen des Sammelbandes sind 7 überwiegend methodisch und 17 Beiträge stärker anwendungsorientiert. Die Palette der aufgezeigten Anwendungsmöglichkeiten ist außerordentlich breit: chemische Industrie/Verfahrenstechnik (6 Beiträge, zur Kohleverflüssigung (1), Reaktordynamik (1), zu komplexen industriellen Prozessen (2) und zu thermoelastischen und -viskoelastischen Problemen (2)); industrielle Verbrennungsprozesse (4 Beiträge); geothermale Probleme (3 Beiträge); Phasenübergangsprozesse (2 Beiträge; frieren, schmelzen, verdampfen); Umströmungsprobleme im Flugzeug- und Strömungsmaschinenbau (2 Beiträge).

Hervorragend ist die im Buch gebotene Präsentation numerischer Ergebnisse. Die in fast allen Beiträgen zur Illustration gegebenen Isolinienbilder und Vektorfelder zeugen von einem hohen Stand der Softwareentwicklung bei der Anwendung digitalgrafischer Methoden.

Berlin D. Prochnow

Jimenez, J. (ed.), The Role of Coherent Structures in Modelling Turbulence and Mixing. Proceedings, Madrid 1980. Lecture Notes in Physics 136. Berlin-Heidelberg-New York, Springer-Verlag 1981. XIII, 393 S., DM 47,–. US $ 24.70. ISBN 3-540-10289-2

In den letzten Jahren richtete sich das Interesse in der Turbulenzforschung verstärkt auf das Finden und die Untersuchung von kohärenten Strukturen in verschiedenen turbulenten Strömungen. Durch diese Zielstellung wurden neue Experimente angeregt und die auf diesem Gebiet vorliegenden Erkenntnisse wurden auch größtenteils mit Hilfe des Experiments gewonnen. In die theoretische Modellierung und Vorausberechnung turbulenter Strömungen, die etwa in der Technik auftreten, haben diese neueren Entwicklungen bisher noch nicht in ähnlich starkem Maße Eingang gefunden.

Das Ziel der vom 25. bis 27. Juni 1980 in Madrid durchgeführten Konferenz war es, Experimentatoren und Theoretiker zusammenzuführen, um vorliegende Ergebnisse und künftige Arbeitsrichtungen zu diskutieren. Die Vorträge waren in 4 Sektionen gegliedert: Theorie (Vorsitzender: P. G. Saffman; 6 Beiträge), Experimentelle Methoden (J. Mathieu; 3 Beiträge), Experimentelle Ergebnisse (H. E. Fiedler; 4 Beiträge), Anwendungen (A. Liñan; 3 Beiträge). Ein Beitrag jeder Sektion, der von dem jeweiligen Vorsitzenden verfaßt oder mitverfaßt ist, enthält auch zusammenfassende Darstellungen über das betreffende Gebiet.

Freie turbulente Scherschichten und Wandgrenzschichten stehen als Untersuchungsgegenstand im Vordergrund. Mehrere Beiträge zeigen die gegenwärtigen rechentechnischen Möglichkeiten bei der Untersuchung der zeitlichen Entwicklung der Strukturen. In den mehr experimentell ausgerichteten Arbeiten kommen wohl sämtliche Versuchstechniken zur Sprache, die gegenwärtig bei der Entdeckung und Beschreibung kohärenter Strukturen benutzt werden (z. B. Sichtbarmachung, Signalauswertung, „conditional sampling", Spektralmethoden usw.). Die meisten Arbeiten enthalten ein sehr umfangreiches Literaturverzeichnis.

Der Tagungsbericht gibt einen ausgezeichneten Überblick über den gegenwärtigen Stand der Untersuchung von kohärenten Strukturen in turbulenten Strömungen. Auch für auf dem Gebiet der Strömungsmechanik tätige Wissenschaftler, die sich mit diesen neueren Entwicklungen erst bekanntmachen wollen, dürfte das Buch von großem Interesse sein.

Berlin G. Seifert

Gokhfeld, D. A. / Cherniavsky, O. F., Limit Analysis of Structures at Thermal Cycling. Alphen aan den Rijn, Sijthoff & Noordhoff 1980, XXVIII, 537 S., ISBN 90-286-0455-3

In der Monographie werden Probleme der Festigkeit von Konstruktionen bei elastisch-plastischem Materialverhalten auf der Grundlage der Theorie der Anpassung (shakedown) behandelt. Diese Theorie stellt eine Verallgemeinerung der Theorie des Grenzgleichgewichts auf zyklische Belastungen dar, die sowohl mechanischer als auch thermischer Natur sein können. Ein großer Teil des Buches ist der ausführlichen Darstellung der Grundlagen der Anpassungstheorie gewidmet. An einfachen Beispielen von Ein- und Zweiparametersystemen werden die möglichen Effekte elastisch-idealplastischer Körper unter zyklischer Belastung (Augenblickskollaps, alternative Plastifizierung, Anpassung) anschaulich erläutert. Die Voraussetzung des elastisch-idealplastischen Materialverhaltens führt zu einer relativen Einfachheit und Anschaulichkeit der Lösungen. Danach werden das statische Theorem von MELAN und das kinematische Theorem von KOITER als die grundlegenden Prinzipien der Anpassungstheorie behandelt und auf generalisierte Variable verallgemeinert. Die Benutzung dieser Prinzipien zur Bestimmung maximaler zulässiger Belastungsamplituden führt in Abhängigkeit von der betrachteten Klasse von Konstruktionen auf Aufgaben der linearen oder nichtlinearen Optimierung. Diese werden sowohl allgemein behandelt als auch für die konkreten Fälle von rotierenden Scheiben, Druckgefäßen, Platten und Schalen unter den charakteristischen Betriebsbedingungen bei Gas- und Dampfturbinen, Kernreaktoren, Luftfahrzeugen und metallurgischen Ausrüstungen gelöst. Weitere Anwendungsgebiete sind isothermische Kontaktprobleme und die fortschreitende Formänderung bei zyklischer thermischer Belastung. Mit der ausführlichen Behandlung der Anwendungen wird von den Autoren eine Brücke geschlagen zwischen der Theorie elastisch-plastischer Körper und dem Aufgabengebiet der praktischen Dimensionierung von Konstruktionen. Das Buch ist damit sowohl für den Theoretiker als auch für den praktischen Ingenieur von Interesse. Spezielle Kenntnisse werden nicht vorausgesetzt; die benötigten mathematischen Methoden werden jeweils in ihren Grundzügen entwickelt.

Berlin G. MAUERSBERGER

Goldstine, H. H., A History of the Calculus of Variations from the 17th through the 19th Century. Studies in the History of Mathematics and Physical Sciences 5. Berlin-Heidelberg-New York, Springer-Verlag 1980. XVIII, 410 S., 66 Abb., DM 88,–. US $ 51.80. ISBN 3-540-90521-9

Diese bemerkenswerte Neuerscheinung gibt einen inhaltsreichen Überblick über die Geschichte der klassischen Theorie der Varationsrechnung im Sinne EULERS „Methodus inveniendi... und ihrer Weiterentwicklung bis zum Beginn des zwanzigsten Jahrhunderts. D. h., das Buch beschränkt sich im wesentlichen in seiner Auswahl auf solche Untersuchungen, die auf die Bestimmung optimaler Lösungskurven bzw. optimaler Funktionen von einer Variablen orientiert sind. Es umfaßt dafür in großer Breite den historisch bedeutungsvollen Zeitraum von P. FERMAT bis D. HILBERT, A. KNESER und C. CARATHÉODORY und damit eigentlich bis zu jener Schwelle, die wir heute gern als Abschluß der klassischen Variationsrechnung bezeichnen, indem wir einschlägige neuere Erkenntnisse des 20. Jahrhunderts – hervorgegangen unter dem Einfluß funktionalanalytischer Methoden unter Einbeziehung allgemeinerer Optimierungsaufgaben – deutlich davon abtrennen mit Begriffen wie „direkte Methoden der Variationsrechnung", „Variationsrechnung mehrfacher Integrale", „Steuerungstheorie" u. a.

Als Rückblick auf eine überaus fruchtbare Wissenschaftsepoche erscheint das Buch somit zu einem sehr geeigneten Zeitpunkt. Außerdem schließt es eine große Informationslücke, da bis auf Enzyklopädie-Artikel und kommentierte Abdrucke bzw. Übersetzungen fundamentaler Arbeiten von Klassikern der Variationsrechnung seit über 100 Jahren eigentlich kein annähernd gleiches Buch verfaßt wurde. Viele Autoren guter Lehrbücher zur Variationsrechnung haben zwar diesen Mangel dadurch zu beheben versucht, daß sie das mit der Gesamtentwicklung der Analysis und Physik eng verflochtene Wachsen der Variationsrechnung im Anhang oder einleitend aufzeigen. Dem historisch interessierten Mathematiker konnten diese Darstellungen freilich schon lange nicht mehr genügen, er wird begeistert zu diesem neuen Buch greifen, das zugleich dem Lernenden eine leicht verständliche Einführung in die Variationsrechnung vermittelt und Interesse für weiterführende Studien zu erwecken vermag.

Das Buch gliedert sich in folgende Abschnitte: Introduction/ FERMAT, NEWTON, LEIBNIZ and the BERNOULLIS/EULER/LAGRANGE and LEGENDRE/JACOBI and His School/WEIERSTRASS/ CLEBSCH, MAYER and OTHERS/HILBERT, KNESER and Others. Es schließt mit einer ziemlich umfassenden Bibliographie mathematisch-historischer Beiträge zur Variationsrechnung.

Die Grundideen der Klassiker werden mit heutigen Worten in sehr klarer Weise dargelegt, wörtliche Zitate werden nur sehr sparsam eingestreut. Das Werk verzichtet darauf, ausführliche Biographien darzulegen, es konzentriert sich auf das mathematische Schaffen und die gegenseitigen Einflüsse der bedeutendsten Förderer der Variationsrechnung.

Das Buch ist nicht nur inhaltlich sondern auch in seiner äußeren Gestalt und Gliederung sehr ansprechend. Lediglich geringfügige Kritiken wären anzubringen, wie z. B. die Tatsache, daß an Stelle des durchweg mit DU BOIS-RAYMOND bezeichneten Tübinger Mathematikers des 20. Jahrhunderts der Name DU BOIS-REYMOND stehen müßte.

Leipzig R. KLÖTZLER

Noltemeier, H., Informatik I, Einführung in Algorithmen und Berechenbarkeit. München-Wien, Carl Hanser Verlag 1981. 148 S., DM 32,–. ISBN 3-446-13388-7

In der vorliegenden Einführung werden ausgehend vom intuitiven Berechenbarkeitsbegriff einige präzise Algorithmenbegriffe entwickelt und hinsichtlich ihrer Leistungsfähigkeit untersucht. Ausgangspunkt ist der endliche Automat und seine Erweiterung zur unbeschränkten über beliebig langen Wörtern arbeitenden Registermaschine. Die Wörter werden als natürliche Zahlen interpretiert, und es werden entsprechende Grundoperationen (Addition, eingeschränkte Subtraktion, Nulltest) eingeführt. Schließlich wird die Existenz einer mit einer Registermaschine nicht berechenbaren Funktion gezeigt. Anschliessend wird die Äquivalenz zu rekursiven Funktionen gezeigt. Dabei wird wie üblich die Simulationsmethode benutzt. Analog wird die Beziehung zu Turingmaschinen hergestellt und ein Ausblick auf einige weitere Algorithmenbegriffe gegeben (Markoffalgorithmen und Produktionssysteme, v. Neumannsche Maschine, Mehrband-Turingmaschine).

Das Buch ist aus einer Vorlesung für Studienanfänger hervorgegangen und aus diesem Grunde mit einem einleitenden Kapitel mit historischen Bemerkungen und einem sehr knappen Abriß mathematischer Grundbegriffe (Logik und Mengenlehre) versehen. Hier könnte eine ausführlichere Erörterung nützlich sein, zumal auch einige Unklarheiten auftreten. Z. B. kann aus der benutzten Definition des kartesischen Produktes nicht auf seine Assoziativität geschlossen werden. Insgesamt bietet das Buch eine lesenswerte Einführung in die Problematik der Algorithmentheorie und vermittelt eine solide Grundlage für die Behandlung vieler theoretischer Probleme der Informationsverarbeitung.

Leipzig K.-H. BACHMANN

Niemann, H., Pattern Analysis. Informations Sciences 4. Berlin-Heidelberg-New York, Springer-Verlag 1981. XIII, 302 S., 112 Abb., DM 74,–. US $ 33.70. ISBN 3-540-10892-4

Im Buch von H. NIEMANN wird „Musteranalyse" in einem weiteren Sinn als in der klassischen Mustererkennung üblich verstanden und insbesondere der Klassifikation gegenübergestellt. Von einem einheitlichen theoretischen Gesichtspunkt wird vor allem die Analyse komplexer Muster (Bilder, Phonogramme, Szenen, Programme) dargestellt – durchgängig von elementaren Vorverarbeitungsmechanismen bis zur abstrakten diskreten Repräsentation durch Graphen, Listenstrukturen oder Grammatiken. Hervorzuheben ist die Einbeziehung der Darstellung von Steuerstrukturen zur Musteranalyse und moderner, in der künstlichen Intelligenz entwickelter Methoden zur Wissensrepräsentation und -verarbeitung (Frames. Produktionssysteme).

Berlin F. WYSOTZKI

Even, S. / Kariv, O. (eds.), Automata, Languages and Programming. Eight Colloquium, Acre (Akko), Israel, 1981. Lecture Notes in Computer Science 115. Berlin-Heidelberg-New York, Springer-Verlag 1981. VIII, 552 S., DM 49,–. US $ 22.30. ISBN 3-540-10843-2

Der Band 115 der Lecture Notes in Computer Science enthält die Manuskripte von 44 auf der o. g. Konferenz gehaltenen Vorträgen (ausgewählt aus 109 eingereichten) und von einem der beiden eingeladenen Vorträge: ULLMAN, A view of directions in relations database theory. Zur Unterstützung der Aktualität der dargestellten Ergebnisse sind einige Beiträge nur vorläufig formuliert bzw. erweiterte Zusammenfassungen.

Die Konferenzserie steht unter der Schirmherrschaft der Europäischen Vereinigung für theoretische Informatik (EATCS), wodurch der Schwerpunkt der behandelten Themen auf den theoretischen Gebieten der Informatik liegt. Unter den Beiträgen sind allerdings mehrere, die einen direkten Praxisbezug haben bzw. haben könnten.

Hauptsächlich behandelte Gebiete, wobei die Einordnung nicht immer leicht ist, mit ausgewählten Beiträgen waren, etwa nach der Zahl der Beiträge geordnet: Berechnungskompliziertheit (Preparata/Vuillemin, Area-time optimal VLSI networks for computing integer multiplication and discrete Fourier transform; Hong/Mehlhorn/Rosenberg, Cost trade offs in graph embeddings, with applications; Engelfriet/File, Passes, sweeps and visits; Ukkonen, On size bounds for deterministic persers; Snir, Proving lower bounds for linear decision trees), formale Sprachen (Blattner/Latteux, Parikh bounded languages; Istrail, Chomsky-Schützenberger representations for families of languages and grammatical types; Courcelle, An axiomatic approach to the Korenjak-Hopcroft algorithms), Programmverifikation (Sharir, Data flow analysis of applicative programs; Bergstra/Tucker, Algebraically specified programming systems and Hoare's logic; Moricoin/Schwartz, Automatic construction of verification condition generators from Hoare logic), spezielle Algorithmen (Schnorr, Refined analysis and improvements on some factoring algorithms; Reif, Minimum $s-t$ cut of a planar undirected networks in $0(n \log^2(n))$ time), Entscheidbarkeitsfragen (Ehrenfeucht, On the (generalized) Post correspondence problem with lists of length 2; Beeri/Vardi, The implication problem for data dependencies).

Außerdem vertreten waren: Automatentheorie, Definition und Semantik von Programmiersprachen, dynamische Logik, Theorie der abstrakten Datentypen, Kryptologie, Datenstrukturen und -banken (Apt/Plotkin, A Cook's tour of countable nondeterminism; Loeckx, Algorithmic specifications of abstract data types; Lieberherr, Uniform complexity and digital signatures; Meyer/Streett/Mirkowska, The deducibility problem in dynamic logic).

Berlin W. Issel

The Programming Language Ada, Reference Manual. Proposed Standard Document, United States Department of Defense. Lecture Notes in Computer Science 106. Berlin-Heidelberg-New York, Springer-Verlag 1981. X, 243 S., DM 16,50. US $ 7.90. ISBN 3-540-10693-6

ADA ist die jüngste universelle Programmiersprache, die sich wahrscheinlich in großem Maßstab durchsetzen wird. Sie wurde im Auftrage des Department of Defense der amerikanischen Regierung im Rahmen eines Wettbewerbs mit zunehmend detaillierteren Anforderungen entwickelt.

Hauptziele der neuen Sprache waren
- breite Anwendbarkeit (Systemprogrammierung, Numerik, Echtzeitaufgaben)
- Beherrschung großer Programmsysteme
- hohe Programmiersicherheit und -effektivität
- Gewährleistung der Programmportabilität für unterschiedliche EDVA.

Zur Erreichung der genannten Hauptziele wurden bei der Definition spezielle Sprachelemente besonders weit entwickelt: Schrittstellenbeschreibung, separate Compilierung von Moduln, Parallelverarbeitung, strenge Typkontrollen, einfaches Parameterkonzept, Einfluß des Programmierers auf das Laufzeitsystem, Möglichkeiten der Abfrage von Eigenschaften des Compilers bzw. Betriebssystems.

Es wurde bewußt nicht Wert auf eine besonders knappe Niederschrift von Programmen gelegt. Dadurch sind sie in gewissem Grade selbstdokumentierend.

ADA baut zwar weitgehend auf existierende Sprachen auf (Pascal, Modula, Simula 67), enthält jedoch auch originelle Lösungen wie in Abhängigkeit von Datentypen generierbare Programmteile, Synchronisation der Parallelarbeit, Behandlung von Ausnahmesituationen. Die Implementation von ADA erfordert eine Reihe neuer bzw. modifizierter Übersetzungstechniken. Die Compiler werden relativ stark mit statischen Kontrollen belastet, die der Programmsicherheit dienen. In mancher Hinsicht (z. B. Ein- und Ausgabe) ist ADA weitgehend offen, wie überhaupt ein Betrieb dieser Sprache erst mit Hilfe strukturierter Programmbibliotheken möglich bzw. sinnvoll ist.

Realisierungen von ADA sind an verschiedenen Stellen in der Welt für die verschiedensten Rechenanlagen im Gange, etwa ab 1983 werden in größerem Maßstab Compiler zur Verfügung stehen. Ob wirklich eine volle Portabilität der Programme erreicht wird, bleibt abzuwarten.

Der vorliegende Bericht ist eine revidierte Fassung des ursprünglichen, auch in ihm bleibt noch eine Reihe von genauen semantischen Interpretationen einiger Sprechelemente offen. Seine Lesbarkeit wird durch das Fehlen vollständiger erläuterter ADA-Programme erschwert.

Berlin F. Grund

Robinson, D. F. / Foulds, L. R., Digraphs: Theory and Techniques. New York-London-Paris, Gordon and Breach Science Publishers 1980. XV, 256 S., US $ 38.50. ISBN 0-677-05470-X

This is a text book designed to present the theory and applications of directed graphs (digraphs) at an elementary level. It has been written for first or second year university students studying mathematics, engineering, business or the social or biological sciences. The book assumes very little mathematical knowledge. Most of the mathematical necessities are included in a preliminary chapter making the book virtually selfcontained.

Chapter 1 is basically an introduction to digraphs and their elementary properties. An application to logic is included. Chapter 2 is concerned with the structure of digraphs, particularly the component structure. Chapter 3 introduces different types of tournaments and their structure from a theoretical point of view. Chapter 5 deals with rooted trees and their applications in integer programming (branch and bound technique). Chapters 6 and 7 present other applications of digraphs in Operations Research, particularly in activity networks. Chapter 8 is a more theoretical development on some of the matrices associated with digraphs. Finally, chapter 9 introduces in basic concepts of nondirected graphs.

The book contains a large number of exercises from diverse areas. These range from drill type to unsolved problems from the research literature. Solutions to selected exercises are provided at the back of the book. A guide to the literature for future study on the subject is added.

From both the didactic-methodic and the scientific point of view the book is written in an excellent way. The authors seem to be very good teachers knowing the abilities and foibles of students in learning graph theory very well. Hence the book is also a good base for preparing introductory lectures in the theory and applications of digraphs.

Mittweida F. Beichelt

Schuon, E. / Wolf, H., Nachrichten-Meßtechnik. Nachrichtentechnik 9. Berlin-Heidelberg-New York, Springer Verlag 1981. XI, 271 S., 155 Abb., DM 52,–. US $ 24.80. ISBN 3-540-10637-5

Der vorliegende Band Nachrichtentechnik 9 ist eine Buchfassung von Vorlesungstexten in Form einer Monografie und gibt einen aktuellen Überblick über Prinzipien, Verfahren und Geräte zur Nachrichtenmeßtechnik.

In 7 Kapiteln werden in anschaulicher Form physikalische und mathematische Zusammenhänge der eingeführten Meßverfahren in der Nachrichtentechnik vorgestellt. Im ersten Kapitel werden Pegel und Dämpfungsmessungen mit der dazugehörigen speziellen Meßtechnik, einschließlich einer Fehlerabschätzung beschrieben, wobei die Fortschritte durch Einsatz sogenannter „intelligenter" Meßgeräte breiten Raum einnehmen. Das zweite Kapite behandelt Phasen und Laufzeitmessungen, besonders auch unter dem Aspekt der digitalen Signalübertragung in der PCM und Datenübertragungstechnik. Kapitel 3 behandelt Probleme des Einschwingverhaltens zur Ermittlung von Systemeigenschaften. Hier werden die besonderen Anforderungen an die Meßtechnik eingehend aufgezeigt. Das 4. Kapitel ist den Impedanzmessungen gewidmet, wobei neben der Absolutmessung des Scheinwiderstandes auf die Bedeutung der Messung des Relfexionsfaktors eingegangen wird. Die Messung nichtlinearer Verzerrungen werden im Kapitel 5 erläutert. Hier wird auch der Versuch gemacht, eine theoretische Behandlung dieses schwierigen Meßkomplexes vorzunehmen und der Einsatz der Rausch-Klirrmeßtechnik im Vergleich zu anderen Meßverfahren (Differenztonmessung, Klirrfaktormessung, Intermodulationsverfahren) gegenübergestellt. Das Kapitel 6 beinhaltet die Messung stochastischer Größen unter Einsatz von Rauschgeneratoren zur sogenannten Korrelationstechnik oder Korrelationsanalyse. Das letzte Kapitel beschreibt Meßmethoden bei der digitalen Datenübertragung und der PCM-Technik. Die Besonderheit der dazugehörigen Meßtechnik und Meßverfahren werden theoretisch und an praktischen Beispielen demonstriert.

Die Aktualität dieses Buches wird deutlich durch Einbeziehung von Meßverfahren an Lichtleitern. Der übersichtliche Aufbau der einzelnen Kapitel und die zahlreichen Literaturhinweise machen das Buch für den Meßtechniker wie auch für den Theoretiker zum Nachschlagewerk und täglichen Arbeitsmittel.

Berlin W. Meschkat

Katok, A. (ed.), Ergodic Theory and Dynamic Systems I. Proceedings, Special Year, Maryland 1979-80. Progress in Mathematics 10. Boston-Basel Stuttgart, Birkhäuser Verlag 1981. XI, 333 S., ISBN 3-7643-3036-8

Die Theorie der dynamischen Systeme ist eines der Gebiete, auf denen sich theoretische Forschung und vielfältige Anwendungen in zunehmendem Maße beeinflussen. Gerade in der letzten Zeit sind hier theoretische Ergebnisse erzielt worden, durch die man hofft, zu Modellen für so komplizierte Vorgänge wie etwa die Turbulenz zu gelangen. Berücksichtigt man, daß zudem neue Methoden entwickelt wurden, die zu interessanten innermathematischen Erkenntnissen geführt haben, so nimmt es nicht wunder, daß hier ein reges Betätigungsfeld für Mathematiker mit der entsprechenden Vielfalt von Publikationen entstanden ist. Dabei beschränken sich die Forschungen nicht auf einen eng begrenzten Teil der Mathematik, sondern benutzen Erkenntnisse aus verschiedenen Gebieten. Schon von Poincarés himmelsmechanischen Untersuchungen her ist die Dynamik eng mit der Maßtheorie (d. h. mit der maßtheoretischen Dynamik oder Ergodentheorie) verbunden, und auch für die später zunächst mehr geometrische Richtung, die durch die qualitative Theorie allgemeiner gewöhnlicher Differentialgleichungen (und z. T. auch partieller Evolutionsgleichungen) motiviert wird, ergeben sich immer wieder Zusammenhänge zur Maßtheorie und statistischen Mechanik.

Das vorliegende Buch besteht aus einer Reihe von sieben Artikeln, die den Charakter von Originalarbeiten haben, von denen sechs der Ergodentheorie zuzurechnen sind, während einer, er ist mit 123 Seiten der überaus längste, einen interessanten Beitrag zur klassischen konservativen Mechanik liefert. Das Buch steht also nicht unter einem einheitlichen engeren Thema, sondern es bringt aktuelle Beiträge, die auf einem der Dynamik und Ergodentheorie gewidmeten Jahr am Mathematischen Institut der Universität von Maryland mit Teilnehmern aus vielen Ländern vorgestellt und diskutiert wurden.

Die Autoren und Titel der Beiträge sind: D. J. Rudolph: A characterization of those processes finitarily isomorphic to a Bernoulli shift. — D. S. Lind: Finitarily splitting skew products. — A. del Junco: Disjointness of measure-preserving transformations, minimal self-joining and category. — A. del Junco/M. Keane/B. Kitchens/B. Marcus/L. Swanson: Continuous homomorphisms of Bernoulli schemes. — W. A. Veech: Projective swiss cheeses and uniquely ergodic interval exchange transformations. — Y. Katznelson/B. Weiss: When all points are recurrent/generic. — R. L. Devaney: Singularities in classical dynamical systems.

Es ist bei dieser Besprechung natürlich nicht möglich (und im Hinblick auf die Referatenorgane wohl auch nicht nötig), die einzelnen Arbeiten zu referieren. Bzgl. der Ergodentheorie sei daher nur vermerkt, daß sich in mehreren der vorgelegten Ergebnisse des Bemühen zeigt, in bisher abstrakt maßtheoretische Zusammenhänge auch die den „Phasenräumen" zugrundeliegende natürliche Topologie einzubeziehen und, daß eine der längeren Arbeiten (die von W. A. Veech) den sowohl wegen der Anwendungen als auch wegen ihrer Theorie so aktuellen maßerhaltenden Abbildungen eines Intervalls auf sich gewidmet ist. In der Arbeit von R. L. Devaney werden Singularitäten eines konservativen mechanischen Systems betrachtet (das sind Punkte im Phasenraum, in denen die Gleichungen — wie etwa bei Kollisionen im n-Körperproblem — nicht definiert sind). Durch eine geometrische Konstruktion gelingt es, das Verhalten der Bahnkurven in der Nähe solcher Singularitäten durch ein einfaches (allerdings nicht mehr hamiltonsches) dynamisches System zu beschreiben. Dieser Artikel scheint die bisher ausführlichste und auch am leichtesten zugängliche Darstellung dieser interessanten Theorie zu sein.

Berlin H. G. Bothe

Schwarz, H., Optimale Regelung und Filterung. Reihe Wissenschaft. Braunschweig-Wiesbaden, Vieweg 1981. 129 S., DM 29,80. ISBN 3-528-03071-2

Der vorliegende Band ist als Fortsetzung des in der gleichen Reihe erschienenen Einführungsbandes „Zeitdiskrete Regelungssysteme" ausgelegt. Behandelt werden Probleme der optimalen Regelung linearer Systeme, zeitdiskrete Filter, stochastische Signale in zeitdiskreten Systemen, Beobachter und optimale Filter. Die Darlegung des Materials ist klar und einfach und auf das Wesen der Methoden konzentriert, die mathematischen Ausdrücke werden schnörkelfrei und in leicht überschaubaren Schritten abgeleitet. Die Ausklammerung der in vielen Lehrbüchern eher ablenkenden Anwendungsbeispiele gibt dem Band eine angenehme Strenge und Geschlossenheit und gestattet dem Autor, auf knapp 120 Seiten die wesentlichsten Ideen und Verfahren der optimalen linearen Regelung und Filterung im zeitdiskreten Fall darzustellen.

Zusammen mit dem Teil „Zeitdiskrete Regelungssysteme" stellt dieser Band eine ausgezeichnete Einführungslektüre für Studenten der Meß- und Regelungstechnik, für an Regelungsproblemen interessierte Mathematiker, Physiker, Informatiker u. a., aber auch eine klar geordnete Nachschlagmöglichkeit für den Spezialisten dar.

Bedauerlich — wie auch im Einführungsband — einige Druckfehler, besonders in den Formeln.

Berlin V. Kempe

Csörgö, M. / Révész, P., Strong Approximations in Probability and Statistics. Budapest, Akadémiai Kiadó 1981. 284 S., Ft 180,—. ISBN 963-05-2441-4

Die fundamentale Aussage des zentralen Grenzwertsatzes der Wahrscheinlichkeitstheorie, wonach z. B. bei einer Folge (X_n) unabhängiger und identisch verteilter Zufallsgrößen mit Mittelwert 0 und Streuung 1 für die Folge (S_n) der Partialsummen $S_n = \sum_1^n X_m$ eine Grenzwertbeziehung $\mathsf{P}\{n^{-1/2} S_n \leqq y\}$ zur standartisierten Normalverteilung $\Phi(y)$ besteht, kann auch dahingehend interpretiert werden, daß die einzelnen Summen S_n die ursprüngliche Verteilung der X_n „asymptotisch vergessen". Andererseits besagt der Hauptsatz der Mathematischen Statistik, daß durch die gesamte Folge (S_n) die Verteilung der X_n mit Wahrscheinlichkeit 1 definiert ist. Betrachtungen dieser Art und die zugehörigen Methoden, zurückgehend insbesondere auf Erdös und Kac, kennzeichnen das sog. Invarianzprinzip. Ein wesentlicher Impuls ergab sich seitens Donsker durch stetige Interpolation der S_n zu einem stochastischen Prozeß $\{S_n(t); 0 \leqq t \leqq 1\}$ auf $C(0, 1)$ und der Konvergenzbeziehung $h(S_n(t)) \to h(W(t))$, gültig in Verteilung, wobei $W(t)$ einen Standard-Wiener-Prozeß und h ein beliebiges stetiges Funktional bezeichnen. Die von Strassen bewiesene Beziehung $|S_n - W(n)| / \sqrt{n \log \log n} \to 0$ ist, wegen der Gültigkeit mit Wahrscheinlichkeit 1, ein Beispiel eines starken Invarianztheorems. Das hauptsächliche Anliegen des vorliegenden Buches bildet nun die Thematik der starken Invarianz in Form der Zusammenstellung sowie Erarbeitung — und dies unter wesentlichem eigenen Anteil der Autoren — von Resultaten für Partialsummen sowie empirische Prozesse — letztere ergeben sich durch Standardisieren aus empirischen Verteilungsfunktionen — sowie der Diskussion über die Anwendbarkeit der Methodik starker Approximation auf Probleme der Wahrscheinlichkeitstheorie und mathematischen Statistik. Der Stoff ist in sieben Kapitel gegliedert. Kapitel 1 beinhaltete über den Wiener-Prozeß sowie über Gaussche Prozesse benötigte Aussagen, die mittels der Methoden der starken Approximation auf Partialsummen und empirische Prozesse erweitert werden können, während Kapitel 2 der Frage bestmöglicher Approximation von Partialsummen durch einen WIENER-Prozeß nachgeht. In Kapitel 3 steht die Frage im Mittelpunkt, wie groß bzw. wie klein die Zuwächse von Partialsummen unabhängiger, identisch verteilter Zufallsgrößen sein können. Kapitel 4 enthält Theoreme über die starke Approximation von empirischen und Quantil-Prozessen durch geeignete GAUSSsche Prozesse. Kapitel 5 behandelt empirische und Quantil-Prozesse unter dem Aspekt der starken Approximationsmethoden, u. a. auch in dem Fall, daß Parameter zu schätzen sind, oder für Tests. Kapitel 6 bezieht sich auf starke Approximation von Dichten sowie Regressions- und charakteristische Funktionen. In Kapitel 7 schließlich wird die Methodik der starken Approximation bei Partialsummen und empirischen Prozessen mit zufälliger Summandenanzahl betrachtet. Insgesamt liefert das Buch durch die Vermittlung neuer und interessanter Aussagen, durch die Darstellung von Methoden und Zusammenhängen und nicht zuletzt durch ausführliche Literaturangaben einen außerordentlich wertvollen Beitrag zu modernen und aktuellen Problemstellungen der Stochastik. Die Kompetenz der Autoren für diese Thematik sowie die Präzision wie auch Eleganz ihrer Darstellung werden dem Buch eine weite Verbreitung und eine hohe Wertschätzung sichern. Konzentriert sich die vorliegende Monographie auf den Fall unabhängiger und identisch verteilter Zufallsgrößen sowie eindimensionale reellwertige Zeit- und Zustandsparameter, so ist die allgemeinere Situation eines Multi-Zeit-Parameters für einen folgenden zweiten Band vorgesehen.

Dresden P. H. MÜLLER

Dallmann, H. / Elster, K.-H., Einführung in die höhere Mathematik für Naturwissenschaftler und Ingenieure. Band 2. Jena, VEB Gustav Fischer Verlag 1981. 580 S., 160 Abb., M 33,–. Ausland M 43,–. Best.-Nr. 5334020

13 Jahre nach dem Erscheinen des ersten Bandes legen die Autoren nun den zweiten Band vor, der zu folgenden Themen Stellung nimmt: Grundbegriffe der Analysis und Topologie, lineare Algebra, Differential- und Integralrechnung reeller Funktionen mehrerer Variabler nebst Vektoranalysis. Das Niveau des eigentlich für Nichtmathematiker bestimmten Buches ist beachtlich und läßt unsere derzeitige Hochschulausbildung weit hinter sich. Es kann daher allen ernsthaften Studenten, auch denjenigen der Fachrichtung Mathematik, als Ergänzungsliteratur empfohlen werden. Hauptunterschiede zu vergleichbaren Büchern sind Abschnitte über Mengenlehre, topologische, metrische und normierte Räume, quadratische Formen, konvexe Funktionen. Man findet eine saubere Abhandlung über implizite Funktionen und die in der heutigen Optimierungspraxis übliche Erweiterung der LAGRANGE-Methode für Extremalprobleme mit Gleichungsrestriktionen, in der auch die Zielfunktion mit einem Parameter multipliziert wird. Zahlreiche sorgfältig ausgewählte Beispiele nebst Übungsaufgaben mit Angabe der Lösung werden dem Studenten das Selbststudium erleichtern. Aber auch der Lehrende wird das Buch mit Genuß lesen, insbesondere, wenn er in der Mathematikausbildung für Naturwissenschaftler und Ingenieure tätig ist. Die Ausstattung des Buches ist vorzüglich. Entgegen ihrer ursprünglichen Absicht wollen die Autoren nun noch einen dritten Band über partielle Differentialgleichungen, Funktionentheorie und Wahrscheinlichkeitsrechnung erarbeiten, auf dessen Erscheinen man mit Freude warten kann.

Dresden H. BIALY

Fahrmeir, L. / Kaufmann, H. / Ost, F., Stochastische Prozesse. Eine Einführung in Theorie und Anwendungen. München-Wien, Carl Hanser Verlag 1981. XII, 366 S., 44 Abb., DM 44,–. ISBN 3-446-13411-5

Im Unterschied zu zahlreichen Grundlagenbüchern über stochastische Prozesse wird in der vorliegenden Arbeit neben einer leicht verständlichen Darstellung der mathematischen Theorie auch auf vielfältige Anwendungsmöglichkeiten in den Naturwissenschaften, der Technik und der Wirtschaft ausführlich eingegangen. Schwerpunkte der Ausführungen sind Markovsche Ketten, diskrete Markovsche Prozesse mit stetigem Parameter, Erneuerungsprozesse, Semimarkovsche Prozesse, stetige Markovsche Prozesse mit stetigem Parameter, stationäre sowie nichtstationäre Prozesse.

Nachdem der Leser mit grundlegenden Definitionen und Eigenschaften der stochastischen Prozesse vertraut gemacht worden ist, erfolgt eine sehr anschauliche Einführung in das Gebiet der Markovschen Ketten. Im Abschnitt über diskrete Markovsche Prozesse wird neben grundlegenden Ausführungen zur Theorie und zu Grenzwertsätzen insbesondere auf die praktisch wichtigen Poissonschen Prozesse und die Geburts- und Todesprozesse eingegangen. Semimarkovsche und Erneuerungsprozesse werden im Hinblick auf Anwendungen im weitverzweigten Gebiet der Zuverlässigkeitsuntersuchungen diskutiert. Ausführlich sind die Eigenschaften des Wienerschen Prozesses (vor allem seiner Trajektorien) und die Verteilungen einiger Funktionale angegeben. Gleiches gilt für Diffusionsprozesse und stochastische Differentialgleichungen. In den Abschnitten über stationäre und nichtstationäre Prozesse erfolgen die Betrachtungen vor allem unter dem Gesichtspunkt der Prozeßstatistik, speziell modernen Methoden der Zeitreihenanalyse und zum Teil auch der Filtertheorie.

Die theoretischen und anwendungsorientierten Ausführungen zu den verschiedenen Klassen stochastischer Prozesse werden durch die Behandlung statistischer Aspekte abgerundet, auch auf praktisch bedeutsame Simulationsverfahren und geeignete Näherungsmethoden zur Berechnung verschiedenster interessierender Größen wird eingegangen. Eine sehr gute Aufgabenzusammenstellung (mit Lösungen) am Ende eines jeden Abschnitts und sorgfältig ausgewählte anschauliche Beispiele dienen dem Verständnis und ermöglichen dem Leser, mit Grundkenntnissen aus der Wahrscheinlichkeitsrechnung, Statistik, Analysis und linearen Algebra den Inhalt im Selbststudium gut zu erfassen und zu verarbeiten.

Berlin C. LANGE

Hughes, P. / Brecht, G., Die Scheinwelt des Paradoxons. Eine kommentierte Anthologie in Wort und Bild. Braunschweig, Vieweg 1978. 120 S., ISBN 3-528-08379-4

Zunächst empfiehlt der Rezensent, nicht seine Rezension, sondern stattdessen diese herzerfrischende Anthologie selbst zu lesen, da er nämlich seine Ansicht über sie für fragwürdig hält, ohne von der Richtigkeit dieser Ansicht überzeugt zu sein. Dieses Buch muß man ganz lesen, wenn man wissen will, was drin steht, denn es gibt vernünftigerweise kein Inhalts- oder Sachverzeichnis, keine Kapiteleinteilung oder ähnliches; dafür aber kann man die einzelnen Beiträge, von einer Zeile bis etwa zwei Seiten lang, in beliebiger Reihenfolge vernaschen, ausgenommen die vielen „weißen Flecken", die aussehen, als habe eine Zensur nicht hinreichend paradoxe Paradoxa gestrichen. Das Buch wendet sich an niemand, nämlich an jedermann, an Geister, die tiefsinnig mit ihrem Geiste an ihrem Geiste arbeiten wollen, wie an solche, die nur ihren Spaß suchen; etwaige Vorbildung wäre unnötig, aber unschädlich; das Buch enthält Beispiele, die Wissenschaftsgeschichte gemacht haben, wie die Paradoxa von ACHILLES und der Schildkröte oder dem Dorfbarbier, weniger bekannte Aussprüche bedeutender Leute — ich zitiere nur die bis heute unbeantwortete Frage: „Was wird aus dem Loch, wenn der Käse alle ist?" —, aber auch solche die nicht mit großen Namen verbunden sind, wie etwa den allgemeinen Satz „Allgemeine Sätze sind einen Dreck wert", der leider schon einem amerikanischen Oberrichter zugeschrieben wird; auch graphische Paradoxa, wie z. B. Illustrationen von ESCHER gibt es zu sehen. Abschließend fragt sich der Rezensent, da er dem Buch weite Verbreitung wünscht, ob er sich nicht weigern sollte, einer Redaktion, die seine Rezension zu drucken bereit ist, seine Rezension zur Verfügung zu stellen.

Berlin W. HEINRICH

Batschelet, E., Einführung in die Mathematik für Biologen. Berlin-Heidelberg-New York, Springer-Verlag 1980. XIII, 557 S., 226 Abb., DM 68,–. US $ 40.00. ISBN 3-540-09633-7

Die Zeiten, da Biologie und Mathematik nichts miteinander zu tun hatten, auch die, da Biologen einen Bogen um die Mathematik machen konnten, sind vorbei. Folgerichtig brauchen wir eine

„Mathematik für Biologen"; das wäre eine, mit der diese „alles" machen könnten, was sie interessiert, ohne das schreckliche Zeug, von dem die Mathematiker dauernd reden, genauer besichtigen zu müssen, und wenn schon, dann ganz schnell, weil man ja Biologe bleiben will. Eine solche Mathematik gibt es zwar nicht, trotzdem aber kann man in dieser Richtung einiges tun. Gerade das will dieses Buch, und es erfüllt an seinem Platz diesen Zweck. Es bietet eine breitangelegte Einführung in die Höhere Mathematik (man darf von dem, was man beim Abitur angeblich alles gewußt hat, ruhig dreiviertel vergessen haben), genauer werden die Elementaren Funktionen, Anfangsgründe der Analysis und der Wahrscheinlichkeitsrechnung, reelle und komplexe Zahlen, dazu etwas Vektorrechnung, Determinanten und Matrizen behandelt, während z. B. Statistik und Rechentechnik ausgespart bleiben. Vorzug des Buches ist dabei, daß die zahlreichen Aufgaben und Beispiele fast durchweg der Biologie entnommen sind.

Neben einigen harmlosen Schnitzern – Beispiel: Die Lösungen der Aufgaben sind nicht vollständig mit „übersetzt" worden; so steht etwa im Text der Aufgabe 13.5.1. K und M wie Knabe und Mädchen, während es in der Lösung B und G wie Boy und Girl heißt –, neben solchen harmlosen Schnitzern also gibt es auch ganz böse Druckfehler – Beispiel: Auf S. 11 steht zu lesen: „5/0 ist sinnlos, da keine Zahl die inverse Bedingung $x \cdot 5 = 0$ erfüllen würde". Gleich einige Zeilen weiter bleibt der Mathematiker an dem Imperativ „Dividiere nie durch Null" hängen, würde er doch stattdessen lieber „Man kann nicht durch Null dividieren" sehen. Dann aber sagt man sich, daß bei einer Diskussion über die Diktion des Textes – wenn schon, gäbe es viele Stellen, über die man reden könnte – ja auch beachtet werden muß, daß ein Biologe und ein Mathematiker ein und denselben Satz anders lesen. Überwiegend werden die Biologen den Text wohl so lesen, wie sie sollen, und also Gewinn aus dem Buch ziehen, sofern sie noch keine Mathematiker sind.

Berlin Redaktion

Weber, L. / Mendoza, E., Kabinett physikalischer Raritäten. Facetten der Physik 1. Braunschweig-Wiesbaden, Vieweg 1980. XIV, 210 S., zahlr. Abb., DM 24,80.

Diese Anthologie ist der erste Band einer Reihe „Facetten der Physik". Der Herausgeber, Roman U. Sexl sagt dazu: „Physik hat viele Facetten: historische, technische, soziale, kulturelle, philosophische und amüsante". Nun, in dieser Anthologie kommen sie alle vor, der gemeinsame Nenner aber ist der letzte Aspekt, der amüsante. Die weit über hundert Miniaturen: Anekdoten, Epigramme, Gedichte, Glossen, Karikaturen, Irrtümer, Zitate usw. wollen zeigen – was manche manchmal nicht mehr so recht glauben –, daß die Wissenschaft Physik nichts Un-Menschliches sei. – Ungewöhnliche Bücher verlangen ungewöhnliche Rezensionen: hat es schon Buchbesprechungen mit Übungsaufgaben gegeben? Wenn nein, hier ist die Premiere: Beweisen Sie, daß der Himmel heißer als die Hölle ist. Hinweis: Man benutze Jesajas 30, 26, das Stefan-Boltzmannsche Gesetz sowie Apokalypse 21,8. Lösung S. 116 des angezeigten Buches.

Berlin W. Heinrich

Manin, Yu. I., Mathematics and Physics. Progress in Physics 3. Boston-Basel-Stuttgart, Birkhäuser Verlag 1981. XII, 99 S., sfr. 22.–. ISBN 3-7643-3027-9

Das vorliegende sehr lesenswerte Büchlein, eine Übersetzung aus dem Russischen, illustriert, wie die Mathematik als Wissenschaft von richtig gedachten idealen Begriffssystemen die gedanklichen Strukturen bereitstellt (oder doch nachträglich zu lichter Klarheit bringt), auf welche die physikalische Erkenntnis Grundzüge der objektiven Realität theoretisch abbildet. Diese Einsicht, daß mathematische Konzepte recht eigentlich die adäquaten begrifflichen Abbilder für die fundamentale Physik sind, wird hier nicht metatheoretisch begründet, sondern durch einige wenige signifikante Beispiele nahegebracht. Das wichtigste ist wohl die mathematische Abstraktion des Raumes, sowohl des linearen als auch des gekrümmten, welche im ersten Kapitel vorgestellt und in der ganzen Abhandlung immer wieder im Zusammenhang mit den verschiedensten physikalischen Theorien, insbesondere modernen, diskutiert wird. Der vierdimensionalen Raumzeit ist dabei ein eigenes Kapitel gewidmet, das sie als Mannigfaltigkeit möglicher klassischer Ereignisse vor Augen führt, als ein physikalisches System somit, aber eben als eines unter vielen. Wie es scheint, ist der Verf. bestrebt, die Sonderstellung des traditionellen, anschaulichen Raumes aufzuheben und ihn in Verbindung mit den abstrakten physikalischen Zustandsräumen zu sehen. So löst er das Zenonsche Paradoxon, wonach der fliegende Pfeil ruht, durch den Hinweis, daß dessen Ort im Dreidimensionalen ja eine Projektion aus dem Phasenraum darstelle. Wir erwähnen indessen dieses Aperçu lediglich als ein Beispiel für die lebendig-einfallsreiche Diktion des Werkes, dessen Schwergewicht vielmehr auf den Generalisationen und den modernen Entsprechungen des klassischen Zustandsbegriffs liegt. Besonders eindrucksvoll ist hierbei die Erläuterung der quantentheoretischen Begriffswelt in ihrem grundsätzlichen Unterschied zur klassischen gelungen. Weitere in der zu besprechenden Schrift beleuchtete Kategorien sind physikalische Größen, Dimensionen und Konstanten, sowie Wirkung und Symmetrie. Selbstverständlich verbietet der geringe Umfang des Büchleins das Eingehen aufs Detail. Umso deutlicher rücken die wesentlichen Gesichtspunkte zusammen, zumal es dem Verf. gegeben ist, sie mit sicherem Blick zu erfassen und in ebenso schlichten wie präzisen Worten auszudrücken. Deshalb ist die Lektüre seiner Zusammenschau von Mathematik und Physik nicht nur für gestandene Fachleute dieser Disziplinen, sondern auch für ihre Studenten erfreulich und nutzbringend.

Dresden H.-G. Schöpf

Gonser, U. (ed.), Mössbauer Spectroscopy II. The Exotic Side of the Method. Topics in Current Physics 25. Berlin-Heidelberg-New York, Springer-Verlag 1981. XII, 196 S., 67 Abb., DM 62,–. US $ 28.20. ISBN 3-540-10519-0

Während sich der erste Topics-Band (Vol. 5), der 1975 veröffentlicht wurde, mit der Anwendung der Gammaresonanz in den klassischen Disziplinen wie Chemie, Magnetismus, Biologie, Geologie, Mineralogie und physikalische Metallurgie befaßt, präsentiert das gegenwärtige Buch intellektuelle Verknüpfungen mit dem klassischen Mößbauereffekt. Es ist ein typisches Beispiel für interdisziplinäre Forschung und dient der Inspiration der Wissenschaft bis zur gegenwärtigen Grenze der Erkenntnis. Die hier beschriebenen Ideen und Anwendungen sind teilweise so ungewöhnlich, daß man mit Recht von der exotischen Seite der Methode sprechen kann. Einige der verschiedenen experimentellen Ansätze wurden bereits im Labor getestet, andere sind noch reine Gedankenexperimente.

Alle Kapitel zeigen, welch breites Anwendungsfeld entstehen kann, wenn neue technische Möglichkeiten klug genutzt werden, um alte Probleme überraschend elegant zu lösen. Obwohl die unterschiedlichen Beiträge sehr eigenständig sind, wie z. B. die Abschnitte über Relativität, Archäologie und Kunst, kann man doch einige Korrelationen finden. Im Kapitel 2 wird die Phasenbestimmung bei der Strukturanalyse biologischer Makromoleküle mit der Gammaresonanz-Streutechnik an ^{57}Fe-Kernen beschrieben, die auch mit Hilfe der Methode der Synchrotronstrahlung (Kap. 5) durchgeführt werden kann. Das gleiche gilt auch für den mößbauerspektroskopischen Nachweis der Rotverschiebung im Gravitationsfeld der Erde (Kap. 3). Im Kapitel 4 wurden die Grundlagen zur Entwicklung eines Gammalasers dargelegt, dessen praktischer Nutzung man jetzt bereits näher gekommen ist. Während bei der Gammastrahlen-Polarimetrie (Kap. 6) mehr theoretische Aspekte im Vordergrund stehen, befaßt sich das Kapitel 7 mit einer praktischen Anwendung der Konversionselektronen-Mößbauerspektroskopie (Implantate).

Obwohl der vorliegende Band an Experten addressiert ist, wendet er sich auch an alle Wissenschaftler, die an der Frage interessiert sind, was man mit der Gammaresonanz-Methode gemacht hat und welche Anwendungsmöglichkeiten sich noch im Zustand der Erkenntnis befinden. Naturgemäß werden deshalb viele Fragen aufgeworfen, die noch einer Lösung harren. Das Buch leistet damit einen excellenten Beitrag zur Vorstellung neuer Aspekte des Mößbauereffektes, wobei fundierte Grundlagenkenntnisse vorausgesetzt werden, und beinhaltet neue Vorstöße in das Gebiet unbekannter Phänomene und Anwendungen der rückstoßfreien Gammaresonanz.

Berlin H. Mehner

Toda, M., Theory of Nonlinear Lattices. Solid-State Sciences 20. Berlin-Heidelberg-New York, Springer-Verlag 1981. X, 205 S., 38 Abb., DM 69,–. US $ 31.40. ISBN 3-540-10224-8

Die Theorie nichtlinearer physikalischer Systeme hat in den letzten Jahren ganz bedeutende Fortschritte erfahren. Eine Nichtlinearität ist dabei gegeben durch einen nichtlinearen Zusammenhang zwischen Ursache (z. B. elastischer Kraft) und Wirkung (z. B. Auslenkung eines Oszillators) bei einem dynamischen System. Diese Fortschritte sind zu einem wesentlichen Teil neuen Erkenntnissen in der Theorie nichtlinearer Differentialgleichungen zu verdanken, die von Mathematikern und Physikern erarbeitet wurden. Der bekannte japanische Theoretiker MORIKAZU TODA hat dazu seit 1966 wichtige Beiträge geliefert, insbesondere durch Entwicklung der Theorie des heute so genannten Toda-Gitters oder Exponentialgitters, bei dem die Kräfte zwischen den Atomen etwa exponentiell von den Atomabständen abhängen.

Das vorliegende Buch, dessen Original 1978 in Japanisch erschienen ist, beschreibt in geschlossener Darstellung die Theorie der Toda-Gitter im engeren Sinne, d. h. unter Zugrundelegung der klassischen (nichtquantenmechanischen) Bewegungsgleichungen. Charakteristisch ist dabei das Auftreten der nichtlinearen Differentialgleichung von KORTEWEG und DE VRIES, die strenge analytische Lösungen ermöglicht, darunter stoßartige Lösungen, die als solitäre Lösungen oder Solitonen eine grundlegende Rolle in der Physik zahlreicher Klassen nichtlinearer Systeme (Gasplasmen, Festkörper, Elementarteilchen u. a.) spielen. Der Autor bringt in fünf Kapiteln eine historische Einführung, Grundlagen der Physik exponentieller Gitter, Lösungsmethoden, Besonderheiten periodischer Systeme und schließlich den Einsatz der HAMILTON-JACOBI-Theorie der klassischen Mechanik. In einer Reihe von Anhängen werden mathematische und physikalische Einzelprobleme diskutiert, die ein vertieftes Verständnis des schwierigen Stoffes ermöglichen. Weitergehende Entwicklungen, z. B. zu Gittern mit Fremdatomen, zweiatomigen Gittern und quantenmechanischen Gittern, bleiben unberücksichtigt.

Das Werk nimmt eine Mittelstellung zwischen Monographie und Lehrbuch ein. Der Text ist relativ breit und gut lesbar, auch für Studenten höherer Semester. Eine Anzahl Übungsaufgaben mit Lösungen, Literaturverzeichnis, Sach- und Autorenregister vervollständigen den Band. Satz, Druck und äußere Ausstattung sind vorzüglich. Der Springer-Reihe in den Festkörperwissenschaften ist damit ein weiterer wichtiger Band hinzugefügt worden, dessen Studium unbedingt zu empfehlen ist.

Dresden G. VOJTA

In der DDR beziehbar im Rahmen des Kontingents.

EINGEGANBENE BÜCHER

Bei der Schriftleitung sind folgende Bücher eingegangen (ausführliche Besprechung bleibt vorbehalten).

Ikeda, N. / Watanabe, S., Stochastic Differential Equations und Diffusion Processes. North-Holland Mathematical Library 24. Amsterdam-New York, North-Holland Publ. Co. 1981. 480 S., US $ 85.25. Dfl 175.00. ISBN 0-444-86172-6

Friedlander, S., An Introduction to the Mathematical Theory of Geophysical Fluid Dynamics. North-Holland Mathematical Studies 41. Amsterdam-New York, North-Holland Publ. Co. 1980. 282 S., US $ 29.50. Dfl. 65.00. ISBN 0-444-86032-0

Axelsson, O. / Frank, L. S. / van der Sluis, A. (eds.), Analytical and Numerical Approaches to Asymptotic Problems in Analysis. North-Holland Mathematics Studies 47. Amsterdam-New York, North-Holland Publ. Co. 1981. XIII, 366 S., US $ 53.75. Dfl. 110.00. ISBN 0-444-86131-9

Chow, P. L. / Kohler, W. E. / Papanicolaou, G. C. (eds.), Multiple Scattering and Waves in Random Media. Proceedings of the U.S. Army Workshop, Blacksburg 1980. Amsterdam-New York, North-Holland Publ. Co. 1981. IV, 260 S., US $ 38.25. Dfl. 90.00. ISBN 0-444-862-80-3

Glowinski, R. / Lions, J. L. (eds.), Computing Methods in Applied Sciences and Engineering. Proceedings, Versailles, France, 1979. Amsterdam-New York, North-Holland Publ. Co. 1980. X, 724 S., US $ 68.25. Dfl. 140.00. ISBN 0-444-86008-8

Koiter, W. T. / Mikhailov, G. K. (eds.), Theory of Shells. Proceedings, IUTAM, Tbilisi, USSR, 1978. Amsterdam-New York, North-Holland Publ. Co. 1980. XII, 702 S., US $ 117.00. Dfl. 240.00. ISBN 0-444-85338-3

Griffiths, P. A. / Morgan, J. W., Rational Homotopy Theory and Differential Forms. Progress in Mathematics 16. Boston-Basel-Stuttgart, Birkhäuser Verlag 1981. XI, 242 S., ISBN 3-7643-3041-4

Treder, H.-J. / Mücket, J.-P., Große kosmische Systeme. Veröffentlichung des Forschungsbereiches Geo- und Kosmoswissenschaften 10. Berlin, Akademie-Verlag 1981. 91 S., M 15,–. BN 7628508 (2156/10)

Butzer, P. L. / Sz.-Nagy, B. / Görlich, E. (eds.), Functional Analysis and Approximation. Proceedings, Oberwolfach 1980. ISNM 60. Basel-Boston-Stuttgart, Birkhäuser Verlag 1981. 482 S., sFr. 94.–. ISBN 3-7643-1212-2

Vogan, D. A., Jr., Representations of Real Reductive Lie Groups. Progress in Mathematics 15. Boston-Basel-Stuttgart, Birkhäuser Verlag 1981. XVII, 754 S., sFr. 78.–. ISBN 3-7643-3037-6

Arbib, M. A. / Kfoury, A. J. / Moll, R. N., A Basis for Theoretical Computer Science. Texts and Monographs in Computer Science. Berlin-Heidelberg-New York, Springer-Verlag 1981. VIII, 220 S., 49 Abb., DM 42,50. US $ 19.80. ISBN 3-540-90573-1

Duijvestijn, A. J. / Lockemann, P. C. (ed.), Trends in Information Processing Systems. Proceedings, Munich 1981. Lecture Notes in Computer Science 123. Berlin-Heidelberg-New York, Springer-Verlag 1981. XI, 349 S., DM 38,–. US $ 17.70. ISBN 3-540-10885-8

Ponter, A. R. S. / Hayhurst, D. R. (eds.), Creep in Structures. IUTAM/UK 1980. Berlin-Heidelberg-New York, Springer-Verlag 1981. XVI, 615 S., 243 S., DM 120,–. US $ 55.90. ISBN 3-540-10596-4

Ebert, K. H. / Deuflhard, P. / Jäger, W. (eds.), Modelling of Chemical Reaction Systems. Proceedings, International Workshop, Heidelberg 1980. Springer Series in Chemical Physics 18. Berlin-Heidelberg-New York, Springer-Verlag 1981. X, 389 S., 163 Abb., DM 72,–. US $ 33.60. ISBN 3-540-10983-8

FSC
www.fsc.org
MIX
Papier | Fördert gute Waldnutzung
FSC® C083411

Zeitfracht Medien GmbH
Ferdinand-Jühlke-Straße 7
99095 Erfurt, Deutschland
produktsicherheit@kolibri360.de